Meaning in the Multiverse:

A SKEPTIC'S GUIDE TO A LOVING COSMOS

By Justin A. Harnish

CONSILIENCE NOW PRESS

Copyright © 2020 by Justin Harnish. All Rights Reserved.

All rights reserved. No part of this publication may be reproduced, distributed, or transmitted in any form or by any means, including photocopying, recording, or other electronic or mechanical methods, without the prior written permission of the publisher, except in the case of brief quotations embodied in critical reviews and certain other noncommercial uses permitted by copyright law. If you would like to do any of the above, please seek permission first by contacting me at http://justinaharnish.com

Published in the United States by
Consilience Now Press
415 East 3900 South
Salt Lake City, UT 84107
http://justinaharnish.com/

Book Layout ©2015 BookDesignTemplates.com
Cover Design ©2020 Jonathan Smith

Ordering Information:
Quantity sales. Special discounts are available on quantity purchases by corporations, associations, and others. For details, contact the "Special Sales Department" at the address above.

Meaning in the Multiverse:
A Skeptics Guide to a Loving Cosmos
By Justin Harnish

Library of Congress Control Number: 2020916268
Print ISBN: 978-1-7356583-0-8
Ebook ISBN: 978-1-7356583-1-5

TABLE OF CONTENTS

Meaning, Existence, and Experience 1
Why Meaning Matters 10
Existence and Experience 18
    Have you ever been experience? 21
    The World-In-Itself 32
    Oneness with Spacetime 40
    Timelessness 44
Metaphysical Hypothesis 58
Materialism 59
    When stuff shimmers: Wave-Particle Duality 62
    Quantum Smallness 64
    Cosmic Largess 71
    Is Dark Matter Still Matter? 74
    Materialistic Meaning 76
Idealism 83
    Qualia of God 85
    Bit by Bit 93
    Schrödinger's Smeared Cat 94
    On the Seventh Day He... Collapsed Wavefunctions 97
    Spooky No More 99
    Bits from Black Holes 101
    Scientific life after death? 105

Process Ontology	112
The Way	113
The Universe in a Funhouse Mirror	116
The Red Pill	125
Read-Write-Optimize	138
Many Worlds	142
Dual-Slit Experiment	143
Parallelism	149
Consciousness from Clones	155
The Hard Problem of Virtualizing Experience	160
Superconscious Computers	166
Meaning from Many Worlds	169
Optimization	176
Flowing with Existence	180
Deliberate Practice	183
Design Thinking	194
Mindful Experiences	200
Heroic Mindfulness	202
A Wide Mind	208
Synthesis	225
Meaning in the Multiverse	226
Reasonableness	227

Distinguishing Between Natural and Supernatural Universal Meaning 230

Science, Skepticism, and Woo 235

The Morality of Meaning in the Multiverse 238

Meaning in the Multiverse 246

Acknowledgements 252

About The Author 255

Endnotes 257

Dedication:

To Samira,

Your love gives my life meaning in every universe.

PART I:

Meaning, Existence, and Experience

IS LIFE MEANINGFUL?

Across time and status, this question has stood out as one demanding an answer. The implications of a negative result—that life is meaningless—is the actual question on our mind and goes unasked because it is too much for most of us to bear. We have struggled and loved, been cared for and parented, solved problems and anxiously awaited results, valiantly expended energy against disorder, planned, failed, succeeded, lathered, rinsed, and repeated. We are sure that at times, in our best moments, our efforts have not gone unnoticed and that they damn well meant something. We burden our poets, snuggle our children and smell their hair, and meditate in solitude to examine our life quest, whose object—as corny as it may seem—is the meaning of life.

We look into the sea of stars and galaxies, dust and distance, and somehow, though we will never get to these places made vivid by the most advanced eyeballs we have ever made for ourselves, our discoveries adhere to our explanations. We predict the makeup of the stars from pinpricks of light, turn back the clock on cataclysms of creation to source what we see in the present (from light emitted in the distant past), and dig into the next problem, of material and energy that is dark to our light sensing instruments. As much as we search and as unlikely as it may seem, we see nothing that resembles the face we see in the mirror or the experience we see in our "mind's eye." How can the stuff of this existence—so foreign, cold, hostile, and far off—contain anything of meaning for us? Is it all just setting the scene?

Some are content that they have made their own meaning, that the lessons they learned and pass to posterity are indicative of the right way to live, a proof-positive path to purpose. But looking both ways before crossing the angry thoroughfare of history should cause us to forever pause on the curb of the present, because this is what every cohort has said, that it has learned from the myriad mistakes of its ancestors, and passes on the true pearls of wisdom to its descendants. As we confront loss and even our own death, our appetite for intellectual dishonesty about the meaningfulness of life increases. We get into a destructive cycle of tightening the safety-blanket swaddle of metaphysics, dreamed up to comfort us—that something out there shines a light in the darkness—a light that shines only for us.

Most people want so badly for life to be meaningful in the grand scheme of things, that they associate themselves with ancient beliefs that claim on insufficient evidence that a supernatural entity is actively making our lives meaningful. Another large swath piece together a more modern belief that *something* "out there" loves us. This meaning of life is top-down and very impressive. If God or the universe is concerned with our purposeful existence, we can rest assured that our deviations from the path will be minimal and that the ends justify the means. We can do what *feels* reverent: church, peyote, burning sage, and claim God's love only for our in-group. So long as enough other

MEANING, EXISTENCE, AND EXPERIENCE

people are feeding the collective cognitive bias, we feel safe being swept along the dark river whose course is unknown but, we are assured, adheres to a larger, willful plan.

Others assume that human experience is separate from existence. Our inner life is not based in anything explainable by future scientists and this subjective specialness is all that is offered on the meaningfulness menu. It is not our carbon that is distinct, but that it *is like something* to be this collection of carbon that offers purpose, not just to us personally, but also poetically—where we act as the eyes of the world for the dead-inside stuff of existence. Once the mind you've filled with as many points of presence winks out, so too does meaning; *your* life is meaningful if mindful, loving, lived in a transcendent state of flow, or in the service of others, but life in general lacks purpose beyond the personal.

One thing the advocates of a solely personal meaning have gotten absolutely right... there is a lot to discover along the borders of experience and existence. If we take as a foundation that experience is meaningful—and this claim is hard to deny to anyone that has loved; had moments of flow in a musical, academic, or sporting performance; or paid attention to the profound mystery of the illumination of subjective experience *itself*—then the place to start investigating a meaning from existence is in the realms where the distinction between the world-as-it-is and the world-as-we-experience-it is most blurred or paradoxical. These are easy enough to find. Our relationship to space, causality, and especially our experience of time are in a complex and often paradoxical relationship with the physical or computational conception of them. An example that we investigate is the experience we have that time flows from one moment to the next, while time's existence is said to be "frozen," another dimension like up, left, or forward in spacetime[1]. As we become introspective of our experiences and take an honest look at explanations for existence like the Many-Worlds Hypothesis of Quantum Mechanics, our inquiries into the nature of a meaningful existence seem more reasonable.

We have been looking in all of the wrong places to find a universal meaning for our lives. We have asked the question all wrong. Instead of searching for meaning in the heavens, we need to first ask ourselves, "what sort of universe would allow for all-natural universal meaning?" This is where metaphysical speculation makes its appearance. Metaphysics has gotten a bum rap. This once august brand of thought—the philosophy of existence beyond which physics is willing to speculate—is now colluded with every sort of happenstance claim of New Age woo and wizardry. We need to reclaim metaphysics from both the woologists and stoner dorm room alike, for it is a critical tool: advancing science beyond the lab bench of experiment to explanation; offering consilience between disciplines in the humanities like philosophy with the sciences, especially physics and neuroscience; and enlivening the layperson's awe of what really lies just over the horizon of science in the speculations that serious scientists cannot (for a well-founded fear of incredulously being called a metaphysician) make. In this book, we untether our metaphysics and on these open seas find that our most creative conceptions of existence are relevant and vibrant, thanks to the advances of theoretical physics and neuroscience.

> *It seems to me what is called for is an exquisite balance between two conflicting needs: the most skeptical scrutiny of all hypotheses that are served up to us and at the same time a great openness to new ideas. Obviously those two modes of thought are in some tension. But if you are able to exercise only one of these modes, whichever one it is, you're in deep trouble.*[2]
>
> Carl Sagan

MEANING, EXISTENCE, AND EXPERIENCE

Where metaphysics claims existence is fundamentally information or computation, there are numerous reasonable avenues that arrive at an all-natural universal meaning. We choke on the monopoly that materialism has on the frame of what is fundamental in the universe and the stranglehold supernatural speculation has on universal meaning, failing at once to be awestruck by the continued grandeur and complexity we discover and by how profoundly these discoveries and theoretical physics' speculations have changed our frame of what is possible from the universe. The best explanations for existence now offer us insights to the mechanisms for our conscious experience, broaching such profound experience-existence interfaces as the oneness of space, the flow of time, and consciousness from unconscious material. Replacing the frame of a mechanistic cosmos with a more up-to-date model of a computational universe offers us all-natural meaning[3].

Our review of one of the most well-subscribed explanation of reality finds us in a multiverse that frustrates even our most profound intuitions with more wonderment than could ever be created by some desert dime store novelist; a multiverse with many natural places to include meaning that neither manipulates existence through pseudoscience nor inundates humanity with a specialness in the cosmos we do not deserve. What we thought of as solely personal meaning—experiences of flow, mindfulness, or other sorts of profundity and optimized well-being—are processes run in existence, optimization programs on a massively parallel quantum computer we call the multiverse. It is clear to me that there are explanations for the profundity of life to be found in the wonders of the cosmos.

We come full circle to explaining how our experiences are more than just information processes going on in our brain, but instead are an important compilation in how the universe understands itself. These computational processes are compiled across near-parallel universes (that we can never hope to perceive) but whose interference and calculations can be readily experimented upon. No experience happens in only one universe, but as a distribution, a wavefunction of experiences across many worlds. Our efforts to attain peak experience may look the same in this universe but meaning is not just our experience of it and is not separate from existence; experience and existence are integrated when looking "end-on" at the multiverse, across many parallel universes.

Man's search for meaning has been using a water witching rod when tools like the Hubble Space Telescope are available. In *Meaning in the Multiverse*, we will take the lens cap off and stare into the true source of human meaning—the dynamic multiverse.

For twenty years, I have been at the forefront of computational science both as an engineer and a technology strategist for the largest domestic semiconductor memory manufacturer. This training has made me skeptical of explanations that are not falsifiable or that do not stand against their detractors and attempt to correct errors in their logic. However, as an industry engineer and strategist and not an academic, I am able to delve into philosophy without suffering the disingenuous but altogether common impacts a hard-science portfolio would take from a cross-genre work like *Meaning in the Multiverse*. As Sagan is quoted as saying above, my hope is to blend my understanding of the scientific work of giants with their metaphysical antecedents in order to create a new framework for an all-natural universal meaning. I do not do this because I have some god-sized hole in my purpose or in order to undercut a scientific institution that I am on the more practical side of, but instead because I am curious about the biggest questions, the best explanations, and the betterment of our species. I am a fan of all of the human knowledge in the endnotes and like the best explanations of Natural Philosophy available to us thanks to Deutsch, Einstein, Harris, and others. My aim is to be additive to crucial theories surrounding both existence and experience.

MEANING, EXISTENCE, AND EXPERIENCE

In the modern world of relative abundance, we search for the apex of Maslow's hierarchy of needs, we search for meaning. We understand our better selves are inside us and we turn to church, friends, or the self-help section to understand the habits that will draw that awesome-us out for her cotillion. We know the path of least resistance is no way to find ourselves; the easy path is a tar pit of life, where the fossilized remains of the selfish and unexamined lives lie bleached and exposed. We learn from our mistakes. Our path to meaning is a twisted maze full of iteration, edits, and reboots. In most instances, we can't just treat the symptom, we must go to the cause, change our diet to enable our five-hour energy, smile and laugh to be loved, and question our beliefs to grow. In the final part of this book on optimization, we will investigate the practices on offer to optimize a meaningful multiverse and the ramifications that all-natural universal meaning has on our morality, technology, and our species.

To deal with life means to abandon one's self to chaos but to retain a belief in meaning. It is a very serious task.[4]
Hermann Hesse

The speculations of all-natural universal meaning presented in this book are sound and the ontologies they are based on progressively gain in explanatory power over *just* a materialistic metaphysics and *just* personal meaning. Many atheists, agnostics, and secularists have never even made an attempt to find a universal meaning since the ones on offer have collided with their ideas on the makeup of existence, while religious believers (especially those in the West) have only recently come to mindful approaches to personal meaning through the trapdoor of these meditation practices' health benefits.

While there is nothing more important than hypothesizing about "why we are here," what could be less approachable than a treatise on meaning? Who actually answers the question, *"why are we here?"* or *"what is the meaning of life?"* Aren't these questions too overburdened with subjectivity, unknowability, speculation, and even the silliness of answers from the likes of Monte Python and Douglas Adams?

> *"All right," said the computer, and settled into silence again. The two men fidgeted. The tension was unbearable.*
>
> *"You're really not going to like it," observed Deep Thought.*
>
> *"Tell us!"*
>
> *"All right," said Deep Thought. "The Answer to the Great Question..."*
>
> *"Yes..!"*
>
> *"Of Life, the Universe and Everything..." said Deep Thought.*
>
> *"Yes...!"*
>
> *"Is..." said Deep Thought, and paused.*
>
> *"Yes...!"*
>
> *"Is..."*
>
> *"Yes...!!!....?"*
>
> *"Forty-two," said Deep Thought, with infinite majesty and calm.*[5]
>
> Douglas Adams

The answer, as always, is not absolute. There is something to be gained by investigating meaning, in setting a basis for meaning in the universe. I set out to write about meaning to help corporations and then individuals build motivational meaning into their lives, but I have found that humanity's historical universal meaning has helped us in ways similar to what a good corporate vision statement or personal purpose does—it motivates us and aligns our priorities to something larger, something necessary. The meaning of human existence matters in the stewardship of survival, our continued social evolution, and becoming the change we want to see in the world.

Whether speculative, spiritualist, or skeptic, an examination of existence and the titration of meaning from its marrow is a journey sure to excite, intrigue, and motivate us as a species to the next level.

CHAPTER 1:

Why Meaning Matters

HOW MUCH DOES INDIVIDUAL meaning matter? What is your purpose worth to you? For many that have accepted a supernatural meaning or assumed the purpose of their tribe, they have been willing to kill or die for what they imagined their meaning to be. For those that construct a metaphysic that assumes that there is no meaning in an individual life, Albert Camus creates cognitive dissonance and (philosophically) challenges those that would negate meaning to consider their individual survival.

> *There is but one truly serious philosophical problem, and that is suicide. Judging whether life is or is not worth living amounts to answering the fundamental question of philosophy.*[6]
>
> Albert Camus

The mortality of both *the unexamined life not being worth living* and *making the ultimate sacrifice* is instructive and hyperbolic. Many have claimed these maxims and many have carried them through to their ultimate end, but for humanity to have survived past its first conversation on our conviction to live our purpose or prove the purposelessness of it all, there must be another way. The middle path of purpose plied, while continuously examined for errant assumptions, enables life and the diversity of meaning. Camus answers his ultimate question by pressing us into the constant process of examining our purpose.

> *It is good for man to judge himself occasionally. He is alone in being able to do so. But it is bad to stop, hard to be satisfied with a single way of seeing, to go without contradiction, perhaps the most subtle of all spiritual forces. The preceding merely defines a way of thinking. But the point is to live.*[7]
> Albert Camus

While our acceptance of constantly examining our individual meaning matters, our responsibility to live, the selfishness of our genes, outweighs Camus' great philosophical question, and the sacrifice of ourselves and future prodigy for a cause. However, evolution's early victory over the mortality at the extremes of meaning and meaninglessness is not the last straw for the historical value of meaning. The flourishing of our genetic information over time has not progressed solely as a numbers game, but has instead employed a sophisticated socialization, an extended phenotype that like the bird's nest, human civilization is a product brought about by our genetic code[8].

Society is formed from the meaning we imagine for our lives. Humanity's conception of the universe's role for us has altered history more than any other philosophical regime including ethics. The core of both our religious and our political systems aligns society to *why we are here* and *what can (practically) be done about it?*

The meme of meaning in each age is the overwhelming priority for society. The history of human civilization is the history of the conflicts between tribal meanings. The meme of why humanity exists overwhelms governments at all levels, from the city-state to the nation-state, and is the trend forcing most historic occurrences. Many civilizations have clashed over meanings derived from the same roots when slight differences in ge-

Meaning in the Multiverse

ography and language create conflict, as with the wars between different tribes of Abraham. However, there are rare transitions where the main source of meaning is extinguished by a new purpose. Most recently, supernatural universal meaning in the form of religion has been broadly displaced by humanism but not without centuries worth of struggle that still continues into the present day. This epochal shift was catalyzed by early information technology, the printing press, and our most recent revolution in IT, the internet, appears to be assisting another shift in meaning, this one on an even more global scale than the rise of humanism.[9]

Meaning manipulates society not only by prioritizing what we think about, but in its ability to motivate action. Unfortunately, the most just meaning has not always won at a conflict of arms. Humanity's historic delusion in not defining a rational basis for universal meaning and the propensity to instead create regressive memes of tribal meaning is a key reason that a modern rational concept of universal meaning is being addressed in this book.

In our ancient history, the unknown dominated our lives, so *ghost fear* developed as our first source of metaphysical meaning. Spirits were held responsible for the boons and ills of early civilizations. Rites and later, entire mythologies, were setup for superstitious purpose. The temples were both a source of community and a sacrificial crime scene; rites not only built structure into an individual life and defined roles in society, but also ascribed everything to omens and entities that were not there. The meaning in prehistoric society did not so much progress as plod along, sacrificing (literally) the cognitive gains from socialization, language, and experimentation with altered consciousness to the will of fickle gods and their ambitious handlers, the priests. Meaning in the great early societies had life and death consequences, but, as will be the case for much of human history, for the wrong reasons.

As meaning was codified, first by mystics and later by institutions of religion, its role in the maintenance of society was weaponized by one tribe against another. Strict adherence to the meme of meaning was required for admittance into the most religious tribes. Ghost fear was still the norm, but as tribes bifurcated and became more diverse, meaning was modernized as early forms of spirituality, or in other cases, a societal appreciation of natural philosophy was innovated. Societies, like the Greeks and Chinese, that contextualized the universe and consciousness as something to be observed, developed spiritual or academic practices that began to formalize the idea of meaning as something different from the multifaceted will of the gods imposed on humans.

This enlightenment improved human knowledge of the *world-in-itself*, driving the gods of ghost fear from the heavens, from medicine, from the purpose of our existence, and into the gaps. Increased knowledge, once applied, led to technology that granted greater time for leisure and artistic, sporting, spiritual, and studious pursuits approaching the pinnacle of Maslow's hierarchy of needs. Our enlightenment heightened the contrast of a universe where humanity was unexceptional in the cosmos with a politics where kings and religious institutions strained to control the meme of meaning with more oppressive writs of God's divine laws.

Meaning as a motivational and prioritization tool in the maintenance of society was made manifest in the advent of liberal democracies. Resistance to a grant of rights or self-governance has caused wars and unrest for all of the history of liberal democracy. What one segment of a population has as a liberty or rule of law is never easily shared and to the disenfranchised there is nothing liberal or democratic about such segregation. Yet, as we have seen in America and in much of the West, the goal of both the enfranchisement of all in the society and the continuous advancement of liberty and self-governance to meet modern needs is both the means and ends of meaning: individual flourishing is motivated by a level playing field and opportunity that increases with effort and societal gains are steadfast so long as the well-being of others (both present and posterity) is not diminished. The meaning in liberty and democracy is always meant to be a synergy of positive sum.

With liberal democracies running capitalist economies, growth and technological sophistication rose in the world. We raced past the steam age, to the age of electricity, to the nuclear age, the space age, and finally, the information age. In every epoch, meaning was manufactured as rapidly as technology and sometimes, our humanity was lost in the process. The meaning manufactured missed the mark, and quarterly growth paved over *The Commons* we were supposed to leave to posterity. No institution had the brake and indeed no brake likely exists.

Whether due to the duress of modern geopolitical tribal competitions or the disregard or incomplete knowledge of the speed limits on the ethical development of the various technologies, the present moment is likely a tipping point where meaning again needs an upgrade to something that prioritizes species-saving solutions and propels us beyond the many bottlenecks we face. Meaning will finally need to separate from the tribal identity politics to motivate and prioritize the solutions the species will need to steward our survival[10].

In his book *Homo Deus*, Yuval Noah Harari shows how intersubjective entities like religions and governmental structures evolve due to changes in technological sophistication. Harari argues compellingly that our technologies, including the networking of everything (the Internet-of-Things) and our ability to alter our genetic makeup, are leading the intersubjective meaning of our age away from a human-centric one that is responsible for the Enlightenment toward a meaning focused on freedom of data and eventually the creation of algorithms to run society, science, politics, transportation, economics, and ultimately providing for our pursuits of heightened conscious states. When humanism overtook theism as the primary intersubjective entity of meaning, philosophers believed they had solved the problem of whether meaning could be universal. In this epoch where we transition away from meaning focused on human well-being, we will again be forced to reckon with the question if meaning is only an intersubjective entity or instead is emergent from the multiverse.

> *Modernity is a deal. All of us sign up to this deal on the day we are born, and it regulates our lives until the day we die. Very few of us can ever rescind or transcend this deal. It shapes our food, our jobs and our dreams, and it decides where we dwell, whom we love and how we pass away... Modernity is a surprisingly simple deal. The entire contract can be summarized in a single phrase: humans agree to give up meaning in exchange for power.[11]*
>
> Yuval Noah Harari

Humanity should remain optimistic. In the modern age, we have had some success in species-supporting insights. Our greatest advances into understanding the universe, the human mind, medicine, and computation have been due in large part to international collaboration in advancing research in science. Science and its skeptical approach to truth-claims is humanity's single most important tool in approaching a universe that commonly outpaces our intuition; however, our creativity and speculative philosophies are important counterbalances that make space for mathematicians and theorists to build rigor into good and novel explanations.

Enrico Fermi famously considered the paradox between the lack of evidence of other intelligent life in the universe and the high probability that life is not unique to Earth. One rationale for this disagreement between observation and probability came to be known as *The Great Filter*[12]. In this scenario, intelligent life is indeed prevalent, but existential threats overwhelm explanations and resources before these societies can populate a galaxy. Our species stands on the precipice of a *Great Filter* of our own making. One nation or hemisphere cannot take our species through this filter, nor can we overcome this challenge using the same level of thinking that created the problem. Our species' unprecedented social conquest of earth is largely thanks to our resourcefulness in the refining and burning of the fossilized remains of extinct species. However, the negative externalities of climate change only now threaten our existence. Our explanation of the fundamental nature of physics and biology have enlightened us and made us healthier, but have also created terrible nuclear weapons and resistant superbugs. We have utilized our intelligence to overcome numerous problems, but risk relinquishing our intellectual hegemony to artificial intelligences we created but did not contain. As a species, we are now in the unenviable position of being the dog that has caught the mail truck, requiring a new paradigm that will keep us out from under the wheel. For our species, the threats are global and chronic, solvable but requiring a motivating shared vision unlike we have had in the past. We have our reason and science to find and implement solutions, but we need a reason to band together as a species, and moreover, we need to recognize that our place in the universe is not determined by how small we are on the cosmic scale or how young we are as a civilization, but by our ability to use our knowledge to solve problems and by our humanity to discover new modes of meaning that band us together.

CHAPTER GLOSSARY

Personal meaning - meaning available to conscious entities from their experiences.

Universal meaning - meaning that is available thanks to the unique makeup of existence. Universal meaning can be caused by either natural or supernatural means.

Supernatural - Actions or entities that are contrary to the Laws of Physics, that is the natural world. While the multiverse might be speculative, it is natural as it is derived from physical principles and quantifiable in mathematics; gods, miracles, and the karmic wheel of rebirth are not physically possible without their being from a separate supernatural realm.

Ontology - a speculation about the way things are. A branch of philosophy also known as metaphysics, what physics is "getting at."

Executing to meaning - Meaning prioritizes the most important ideas in society and motivates actions. Societies collide when their underlying intersubjective entities like meaning are out of sync.

CHAPTER 2:

Existence and Experience

CONCEPTS
- We investigate the distinction between existence and our experience of it.
- We'll turn internally to consciousness, understanding how actively noticing the fundamentals of our subjective world is a source of insight into the nature of self and experience.
- We'll start the first of our many forays into physical science that sometimes confirms our intuitions of existence but that more often requires creative formulations and mathematics to guide our explanation.
- Reviewing the interrelatedness of everything with space and precisely pondering our experience of the flow of time shapes our strategy through the rest of the book where we will reframe existence and experience in new ways to uncover meaning.

EXISTENCE AND EXPERIENCE

IF WE ARE TO search for the meaning in existence, it is important we understand what existence is. If we are to discuss our experience of meaning, we should be clear on the importance and uniqueness of the life of the mind. From pondering the paradoxes arising as we smear existence and experience, universal meaning arises.

Our esoteric search for universal meaning must be stated simply. We are looking for meaning directed from existence, not only from our experience of it. No supernatural influences will be allowed into our meaning from existence, our explanations of the universe's underlying makeup need to be consistent with scientific models. Existence is not just what is empirical, we will be speculating from the firm basis of the explanations from scientists like David Deutsch and Nick Bostrom. Natural Philosophy is the chisel we will take to the edifice of modern physics to carve out a multiverse of meaning.

Existence is you and me, our neurons, atoms, and quarks... all matter and energy... space and time and spacetime... information and computation. Existence is all-natural! Existence is a trickster: most of matter is empty space; space is likely discontinuous, not a sheet but a web; time does not flow but is frozen; and the laws of physics are not discovered but, are—in fact—the software running the program of existence. We should not expect that the true nature of existence is easy to unlock with our savannah savvy; that we can comprehend the multiverse at all is itself one of the most compelling mysteries for us to ponder.

Since the early parts of the twentieth century, our understanding of existence—the world now explained by quantum gravity, chemistry, biology, mathematics—has undergone a paradigm shift. Before the modern age, much of our understanding of the natural world was based on our direct experience with existence and the intuitions we drew that told us how the world was supposed to work from our limited perspective. We would make deductions about what we experienced with our senses or instruments. Time worked like a clock, light traveled through aether like sound traveled through air, and gravity was just the tensile force of masses working on one another. But now, we leave our intuitions behind, trust in the math and our method, and as a result, our knowledge of existence has grown.

While our intuitions may not be up to the task, our scientific knowledge of the universe objectively and quantitatively proves that the laws governing existence are comprehensible. Existence agrees with our best explanations to precisions of one part in one billion. In most domains of applicability, existence is measured and managed for human comfort and homeostasis with ubiquitous gadgets like the thermometer and natural gas furnace on a programmable logic controller.

In the modern age when we consider existence, we talk about the universe. From its largest expanses to its smallest fundamental particle and at every gradient in-between, the universe is more well-explained each day. We do not count on magic or more magnificent creatures than ourselves to accomplish this feat, but the progression of our steady and skeptical science that insists on explanations that agree with the way the world works.

Our knowledge of existence has stopped *just* being a means for us to understand it and has become an important component of its composition. As will continue to be a reoccurring theme in this book, our universal machines (both the classical and quantum computers) and our exploration of the cosmos are profound developments, not only for such seemingly insignificant beings as ourselves, but also for the universe as a whole. The only barrier to any technology allowed by the laws of physics and the resources of the universe is the knowledge to design and build it. One example of this is that the coldest place in the known universe is in Boulder, Colorado. Here, a team of physicists and engineers created a machine that reliably reduces the temperature in its hold to a few hundredths of a degree Kelvin above absolute zero.

EXISTENCE AND EXPERIENCE

Richard Feynman, the Nobel Prize winning physicist, relates a story about himself and his artist friend that illustrates how our understanding of existence enhances our experience of it.

Doctor Feynman's artist friend pitied him for not having a true experience of the beauty of a flower. The artist felt that his appreciation of the flower was more profound, that his mimicry of the flower in art revealed his mastery of the abstraction of beauty and the nature of the flower as a thing-in-itself. He accused Feynman of only being able to dissect the flower into mechanisms and "nuts and bolts" components that were soulless.

Feynman's full-throated defense bears witness to the reverence the speculative mind has for all of existence. With the grace of a man whose first language is the mathematics that most simply explain reality, he explains the molecular nature of the flower not as a dissection, but as a dance.

> *At the same time, I see much more about the flower than he sees. I can imagine the cells in there, the complicated actions inside which also have a beauty. I mean it's not just beauty at this dimension of one centimeter, there is also a beauty at a smaller dimension, the inner structure. Also, the processes, the fact that the colors in the flower evolved in order to attract insects to pollinate it is interesting — it means that insects can see the color.*[13]
>
> Richard Feynman

Have you ever been experience?

Consciousness is the mysterious landscape where experience happens. Consciousness forms the context for its contents including, but not limited to, thoughts, emotions, and sensory stimulus. Our subjective experience is our most lucid set of information about existence. Our consciousness is at once the construct for our felt, first-person experience; the space for our inner narrative of thoughts, memories, aspirations, and emotions; and a complex, multidimensional, and interactive model of the phenomena of the outer world. As Sam Harris reminds those of us on his *Waking Up* app and *Making Sense* podcast, "as a matter of experience, there is nothing but consciousness and its contents."

Consciousness is then the subjective, first-person *felt experience*. According to Thomas Nagel's influential paper on the subject, an entity is conscious, if, when we trade places with that entity, it *is like something* to be that entity. If the lights are on, from the inside, then that entity is conscious. While it may be very different being a bat (the use of sonar, the late nights, etc.), so long as trading places does not result in annihilation, the bat is a conscious creature. There are philosophical tricks that we can use to investigate the level of conscious awareness of an entity, but to date, there is no algorithm to objectively prove subjectivity—we only know about consciousness at all because our lights are on.

Even if we are plugged into a simulation like *The Matrix* and this overindulgent computational universe is deluding us with false physical and metaphysical explanations, the one thing that cannot be an illusion is our first-person experience of this simulation. We can only be 100-percent certain of our own consciousness—the subjective nature of consciousness is exactly the point. Not only is consciousness the first dry land free from the evil demiurge; but experience is enlightened-humanity's source of personal meaning, much of its spirituality, and our societal morality. The fact that it is like something to be you and that this subjective experience can be made distinct from thoughts of self, emotions like anger, and can cause profundity or suffering makes it a vital part of an examined life. Compassion is my second-person desire and the agency I take to alleviate your suffering; love is my second-person desire and the actions I take to see you to heightened levels of well-being. Compassion is not empathy, instead of acting as a sink for suffering, we accept suffering as part of the human condition and act in ways that help and are directed by the object of our loving-kindness. These are always positive sum—it does not diminish my mental or emotional capacity to act with loving-kindness—indeed it builds character!

The mystery of the source and distinct evolutionary modules of consciousness is complicated by the difficulty of studying subjectivity through objective (scientific) means. Advances in our understanding of different evolutionary brain modules and their role in different components of consciousness has been accomplished by savvy scientists that merge brain scans with pointed behavioral tests. Components revealed include the lowest level of experience, the body scan, the mental map of sensations from both the musculoskeletal and internal viscera; our feelings like disgust, pain, or pleasure that are not quite emotions but that motivate us like no unconscious thought could; and at the highest level, our narrative and subjective knowledge of experiences throughout time. Once our body maps were in place to detect disease and injury, the modular mind seems to have evolved feelings to motivate our ancestors to *do something about it*, and eventually the narrative sense of self in time evolved to make us more relatable to others[14].

First, and most elusive, is the idea of the construct of consciousness. Like the construct of *The Matrix*, consciousness is the program we are all running that we load everything else into. But consciousness is more than just the backdrop to the contents, it is uniquely *our* experience. Consciousness is a first-person experience of the world that while describable, and to some measure, repeatable, it belongs solely to us. As the thought experiment of "Mary in the Black and White Room" challenges, it is extremely difficult to intellectualize the subjective experience, it seems to exist as a different type of information processing, a different way of "knowing" existence rather than *just* knowing the laws of physics.

> *Mary is a brilliant scientist who is, for whatever reason, forced to investigate the world from a black and white room via a black and white television monitor. She specializes in the neurophysiology of vision and acquires, let us suppose, all the physical information there is to obtain about what goes on when we see ripe tomatoes, or the sky, and uses terms like "red", "blue", and so on. She discovers, for example, just which wavelength combinations from the sky stimulate the retina, and exactly how this produces via the central nervous system the contraction of the vocal cords and expulsion of air from the lungs that results in the uttering of the sentence "The sky is blue." [...] What will happen when Mary is released from her black and white room or is given a color television monitor? Will she learn anything or not?*[15]
>
> Frank Jackson

It may be helpful to build this subjective experience up from its fundamental building blocks in order to get a handle on how it works. For this we'll turn to the work of Antonio Damasio on the components of consciousness that were utilized by the evolving modular brain to build the construct of consciousness: body maps, feelings, and our inner narrative history of ourselves.

For even the beginning meditator, the body map has great utility. It can be used to separate yourself from your ego and see different sensations in the body as just one part of the construct of consciousness. But according to Damasio, the body map was the first sensation the brainstem made its own. Body maps are something more than *just* knowledge. This deepening experience of both the body's external state (essential in the expression of motion critical to our hunter-gatherer ancestors) and of inner sensations (useful to detect health issues before they became acute) was a likely evolutionary advantage. Getting in touch with the sensations of the body has shown strong activation in the brain stem of experienced meditators and a detailed study of this brain activity is consistent with image mapping like is modeled in the more visual regions of the brain. This earliest consciousness is still readily available to mindfulness practitioners today as a way of deeply *embodying* the construct of consciousness.

Next on the evolutionary pathway is the *felt* nature of experiences in consciousness. The difference between the body map and the felt experience of discomfort or pleasure is understandably difficult to distinguish thanks to our more integrated construct of consciousness and the inexactness of words. Sufficed to say, the idea of a feeling reporting to the early components of our brain would have been an evolutionary leap toward the type of subjectivity we experience. Feelings are not emotions; they are the precursors to emotions. We might feel an upset in our stomach or a gag reflex in the internal viscera just under our tongue well before the emotion of disgust arises, our lungs might tighten and our chest internally feels like it takes on a greater weight before we register the attack of anxiety. We take feelings to mean *felt experience*. This idea ties to Nagel's description of the potentialities of consciousness in a bat. We can restate Nagel as "if it *feels* like something to be a bat" without diminishing the nature of the thought experiment. In other words, what it is like to be a bat is to have echolocated your body and *feel like that is you*. We would not expect your *batness* to be anxious about the oncoming dawn!

The construct of consciousness is that of felt experience. Our felt experience is interrelated to the contents of existence and our thoughts, memories, and emotions. We experience it and feel it, personally.[16]

Thanks to the cognitively advanced nature of the cortex—the next module to ascend to a place of preeminence in the workings of consciousness—it is not surprising that felt experience is enriched into a subjective narrative through time. Unchecked, our conscious narrative can separate us from experience, giving us a lot more to think about, but also offering a richer sense of ourselves in the universe and society. This narrative broadens our horizons to where we can see ourselves not simply in the experiences of the here-and-now, but also in those of the hoped-for-future or remodeling our past experiences, internally, to learn from them. We start to feel confident not just in our ability to perform in the present, but to grow and improve in the future. Our narrative sense of self projects us, not just through space and time, but also into society, with our latest evolution in consciousness coming as tribe sizes increased and the early game of *The SIMs* was booting up.

As the self came to mind, our ancestors utilized the idea that they were driving their character in the game of *The SIMs* around from a control room in their head to advance in society, portraying themselves as essential components in the myths of the day. The winners in early society collected the most constituents to their avatar of themselves—this was the meaning to be found in early feudal societies—matching the inner narrative of the importance of self to the outer narrative of the myths of society of the day. In some respects, this continues today with the famous and wealthy sharing a public persona that is far more glamorous than what the reality of their lives are like.

However, the addition of the inner narrative of self has gotten away from us. As Sam Harris states, "feeling like a self is to be thinking without being at all aware that you are thinking." This mindlessness is a tremendous barrier to our ability to appreciate the nature of our thoughts, emotions, and the construct of consciousness.

Sam Harris entitled his book on spirituality *Waking Up* to reveal the selfless nature of consciousness. The analogy he uses to explain our *waking up* to the surface selfless nature of consciousness is of seeing your reflection while staring out a window.

We've all had the experience of looking through a window and suddenly noticing our own reflection in the glass. At that

> *moment, we have a choice: to use the window as a window and see the world beyond, or to use it as a mirror. It is extraordinarily easy to shift back and forth between these two views but impossible to truly focus on both simultaneously. This shift offers a very good analogy both for what it is like to recognize the illusoriness of the self for the first time and for why it can take so long to do it.*[17]
>
> Sam Harris

In *Waking Up,* Harris explores the illusion of the control panel run by 'I' through the medical literature on corpus callosotomy, the separation of the two hemispheres of the brain. After this surgery, psychological continuity is split between the two hemispheres. Any functions that are independently running on just one hemisphere are, upon combination, remembered in the present stream of consciousness.

> *Subjectively speaking, the only thing that actually exists is consciousness and its contents. And the only thing relevant to the question of personal identity is psychological continuity from one moment to the next.*[18]
>
> Sam Harris

The most apparent value of nurturing a selfless state is a liberation from the constant grasping of thoughts and feelings which cause experience to be unsatisfactory — it stresses us out. Rooting yourself in the present context of consciousness creates the conditions for the greatest well-being, the least unsatisfactoriness. You will be more aware of negative emotions or inconsistent logic rising up in consciousness, you will actively notice more of the world and become familiar with your blind spots, and your relationships will not just benefit from more doubt reduction, but also more of a loving acceptance.

If the mindful state of *egolessness* and selflessness was difficult in ancient times, we must be extra tenacious in our modern-day practice since our phenomenal world is one where the ego can be satiated by external inputs of our choosing during every waking moment. At the very moment of writing this line, I have just eaten a warm meal, feel the softness and warmth of cashmere, and am listening to a compilation of classical music of my choosing. This after putting down the audio book I am currently listening to on the fall of Rome. With just a smartphone, I have the potential for constant social interaction, information, and entertainment. The world I make for myself in this way is highly customized but often, nearly completely mindless. Even though I am making every attempt to write about the acute sense of full-mindedness to experience, I am distracted and purposefully distracting myself in an effort to accomplish a task of description that is limited to one person—to me. As Hofstadter so eloquently states, the truth of the selflessness of consciousness and my unique subjective worldview are an intriguing pairing of opposites.

> *In the end, we are self-perceiving, self-inventing, locked-in mirages that are little miracles of self-reference.*[19]
>
> Douglas Hofstadter

As anyone who has attempted to be mindful of experience knows, the short temporal nature of the experience is frustrating. The self bears down on the present moment with a spectrum of thought including everything from detritus to the essential, from the timely to the random. Experienced meditators do not hope to eliminate the errant thought, but instead to become aware of its arrival and exit and nothing more. The stream of consciousness is an apt metaphor and it is the meditator's job to, over time, travel from the tumultuous headwaters to the delta where the muddy silt of the self can settle, and a clear stream can rejoin the source of all experience.

EXISTENCE AND EXPERIENCE

A thought process used to engage the selfless subjective construct is that of finding of your own *headlessness*. *Headlessness* is a profound meditative approach of actively noticing the nature of the construct of consciousness. Headless meditation was the adroit observation of Douglas Harding upon seeing a subjective self-portrait drawn by Austrian physicist and philosopher Ernst Mach.

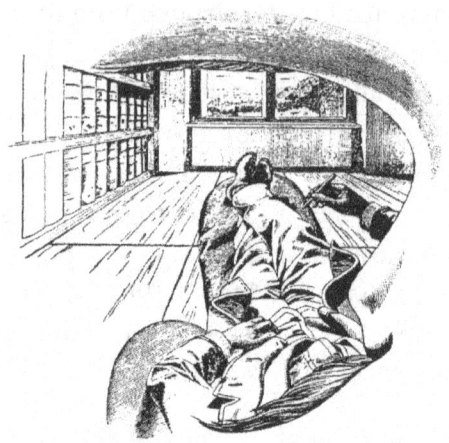

Fig. 1 - Subjective self-portrait drawn by Austrian physicist and philosopher Ernst Mach

Most simply, you can direct your attention out into the world or down onto yourself, and depending on the various calibrations of your senses and the conditions of the environment, you will see the scene around you and some part of your familiar body. However, if you direct your attention to your head, you'll notice that what is there is not really your head, but the whole world in-view and your internal thoughts. Obviously, you cannot see your head (except, under some duress to overcome your model, the edge of your nose) without a mirrored surface, but even if you close your eyes, the content of the world you are experiencing, a remembrance or projection into the future, a feeling or a deep thought, is occurring in your head. It is as if, as Douglas Harding states in *On Having No Head,* "your head has been replaced by your experience of the world."[20]

Unlike many of our forms of predictive and pattern-recognition intelligence, an understanding of the neural correlates of consciousness remains elusive. This is known as the *hard problem of consciousness* that roughly states that subjective states do not appear to be constructed from the bottom up, or from some more fundamental informational processing mechanisms. There is no one center nor an easy algorithm for even the most fundamental unit of experience also known as qualia (a *bit* or *particle* of conscious experience). Subjectivity is mysterious and, like was conveyed in the movie *Ex Machina,* hard to test for. If we did not ourselves have a first-person sense of the world, we wouldn't have any sense of what we were missing out on and what to look for. While memory, emotions, and even advanced reasoning logic like is required in math or chess can be isolated and replicated—even optimized—in silicon-based intelligence, thus far the algorithm of consciousness—that it is *like something* to be a conscious thing—is as objectively unknown as it is subjectively present.

> *Arranging atoms in certain ways appears to bring about an experience of being that very collection of atoms. This is undoubtedly one of the deepest mysteries given to us to contemplate.*[21]
>
> Sam Harris

We are a long way off from an algorithm that can model a simple consciousness. A paradox of information science is that modeling those things we find challenging, like chess, is relatively straight-forward for a brute force simulation, but for those things we find easy, even things we handle unconsciously, like motion, modeling quickly becomes complex. Subjective experience is still best interrogated individually.

EXISTENCE AND EXPERIENCE

We are the local embodiment of a Cosmos grown to self-awareness. We have begun to contemplate our origins: star-stuff pondering the stars; organized assemblages of ten billion billion billion atoms considering the evolution of atoms; tracing the long journey by which, here at least, consciousness arose.[22]

Carl Sagan

The traditional way to get in touch with the fundamental nature of experience is meditation. Disciplines each have their own meditation practice. If the goal is to investigate experience, the practitioner will need to increase their concentration and just be present to experience as it fundamentally is. This practice, known as mindfulness or vipassana, builds awareness of the fundamental nature of consciousness by just observing it. It is much harder than it sounds, the mind is constantly introducing claptrap, frustrating novice and experienced meditators alike, causing all of us to wonder "who am I talking to" and "why is it so mundane." However, more experienced meditators will recognize the claptrap as content, just another item in consciousness to use to hone awareness and ultimately be at peace with both the calm and the storm.

There is extraordinary profundity and meaning available from a more mindful experience. Given consistency and good instruction, mindful meditation can help you experience selfless, timeless, and even transcendent states. Much of the meaning that humanity has found to be resilient across generations, that increases our awe and appreciation of the natural world, that brings us presence at work to excel and at home to love most fully, is the result of a more robust awareness of experience, living each moment to its fullest, appreciating whatever appears, good or bad. The meaningfulness we have found from our pharmacological or meditative experiments into experience has done nothing to dampen our historic proclivity of adding religiosity to the investigation of subjective states; however, this is far from necessary. If you are the type of person that sees the Virgin Mary in a piece of toast, the most honest assessment of your experience in prayer is likely to be that you did have a profound experience *and* brought what you to believe to be the proximate cause along for the ride. What is "real" in that moment of flow or mindful experience—and especially the question of whether there is something other than your brain in *existence* also responsible for this experience—is the not-especially-intuitive landscape we set out to chart in this book.

An early vista of the paradoxical landscape we set out to explore occurs to us when we use the words existence and experience and ask the question: does existence have experience? Is your experience of existence the *rare way in which the universe gets to view and appreciate itself?* In our appreciation of consciousness, we have gone from an inconsequential speck on planetary life-support, to offering insight and a unique perspective to the whole of the natural universe. We will continue to uncover meaningful mergers such as poetic naturalism by focusing on the paradoxes that exist on the horizon between existence and experience. Exploring our felt experience of light, energy, space, and time will lead us toward insights essential to unearthing other natural forms of meaning.

The World-In-Itself

Existence is most well-defined in the mathematical explanations of physics. According to physicist Brian Cox of BBC 4's *The Infinite Monkey Cage,* the simplified notation of the equations most fundamental to the makeup of the universe could fit on a postage stamp. These equations include the notation that defines the general theory of relativity, quantum field theory, and entropy. With these equations and their derivations, the forces of the universe, space and time, and the makeup of material can be accurately predicted. Adding information theory further develops how data, memory, and intelligence are developed from the fundamental unit of a bit. What is empirically known about existence is dominated by the geometry of spacetime and the processes of quanta—both of which owe their revelations to the same man—Albert Einstein.

Einstein's special theory of relativity actually confirms our experiences of the world—that free-fall due to gravity appears to have the same weightless quality as one experiences in space. The reverse is also true, that the same g-forces will be present on the human body if standing at sea level or accelerating in space at 9.8 m/s^2.

> *[We] assume the complete physical equivalence of a gravitational field and a corresponding acceleration of the reference system.*[23]
>
> Albert Einstein

The force we feel is the acceleration of our frame of reference—the earth—coming *up* to meet us at a consistent rate change of speed. This constant acceleration of a reference system is accomplished by the unseen curvature of spacetime.

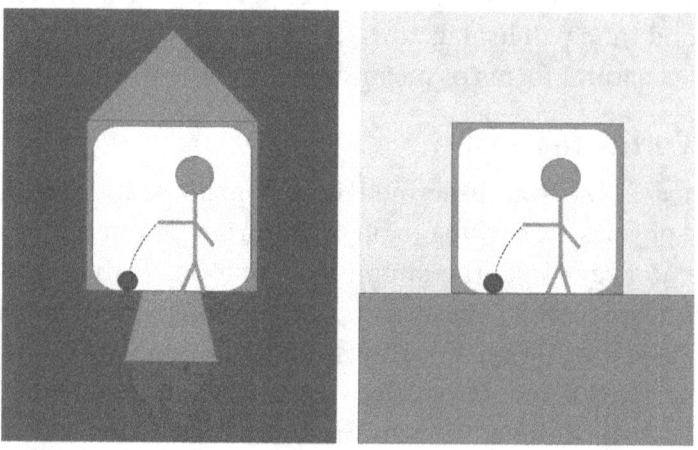

Fig. 2 - Equivalence of acceleration to the force of gravity

General relativity added the curvature of spacetime to special relativity and placed the world as we experience it forever at odds with the actual mechanisms of even the most mundane constituents of existence like light, space, and time. It reframed the universe at planetary scales from one of consistency in time and locality in space to a landscape curving under massive objects, warping to accommodate light speed, and inextricably linked with time.

In order to maintain the consistency of the speed of light, measured as the time-rate-of-change in spatial position, both space and time had to change relative to near-speed-of-light bodies. While not noticeable at our slow day-to-day rates of speed, at the upper-bound of velocity, the speed of light, time and space change to accommodate the constant speed of light. Unlike how two cars traveling in the same direction at the same speed appear to each driver to be standing still, light traveling alongside light would *appear* to be rushing along at the speed of light because space and time would be warped to make it so! This property of spacetime makes it the only medium light could possibly travel upon.

> *How something is, or what its state is, is an illusion. It may be a useful illusion for some purposes, but if we want to think fundamentally, we must not lose sight of the essential fact that 'is' is an illusion.*[24]
>
> Lee Smolin

Not only do both space and time wrinkle to keep the speed of light constant, but spacetime also curves under the mass of cosmic objects like planets and stars. It is this curvature that led theoretical physicists to derive black holes, the big bang beginning, the amount of dark energy required to keep our universe in its current state of expansive acceleration, and the amount of dark matter required to keep our galaxies spinning and together.

EXISTENCE AND EXPERIENCE

What is our experience of spacetime? Of its curvature that creates gravity? Our observations of the curved nature of spacetime around large planetary objects is limited to astronomy. We have seen the impact of gravitational lensing, the curvature of the light of a distant star around our eclipsed sun, and this agrees with the displacement of light's path caused by the curvature we'd expect given the mass of our sun. Even though all massive objects distort spacetime, our experience of spacetime is less entailed with its gravity-creating curvature than in its being the context of existence.

Everything in existence, all matter, both dark and regular, exists in this context. The entire life of each atom in your body and in the room with you will be lived out in the confines of spacetime. Like much of what we will discuss, spacetime is an approximation, a way we can understand the context of material existence, and that we can rest assured has a firm mathematical description. We know spacetime better, not as a couple—instead as individuals: the three dimensions of space and the flow of time. Our experience smeared over the "real world" existence of space and time offers us further insights into natural meaning.

First let us consider space. It's 3-D. In order to dunk a basketball, I need to—let's say—move the ball forward a few inches, maybe left a couple inches, and (much less likely) up at least four feet. I can look out into my surroundings and given the particular, well-modeled combinations of shadow and light and the forms they interact with (chairs, grass, other people) know if I can throw a ball that far. Our visual cortex does a great job of using inputs from our offset eyes to give depth to 2-D representations of 3-D objects, unconsciously turning rod and cone stimulations on the retina into a three-dimensional mind's eye.

We are not nearly as comfortable contemplating higher spatial dimensions.

Mathematically, we can solve for spatial dimensions greater than those we can perceive, all the way out to n-space. Our spatial limitations are clearly shown in the thought experiment, popularized in the book *Flatland* by Edwin Abbott, of what a 2-D individual, A. Square, would see when a 3-D spherical-*deity* passed through Flatland: the circles of its latitude in progression from top to bottom. As *Cube-Landers*, we can imagine how to *fold* a 3-D cube *up and out* of a 2-D segmented cross, but we struggle and fail in our ability to use the same *folding* topological functions to make a 3-D segmented cross into a 4-D *hypercube or tesseract*. *Uber-Up* does not exist for us.

Fig. 3 - Crucifixion (Corpus Hypercubus) by Salvador Dali. 1954

Our perception of space as having only three spatial dimensions limits our ability to know existence. Superstring theory, the most formalized of the general unified theories trying to marry the progress of science of the very small (quantum) with the very large (relativity), utilizes extra dimensions up to a whop-

EXISTENCE AND EXPERIENCE

ping twenty-six to become supersymmetrical. Try as we might, our limited 3-D brains could never overcome their spatial limitations to imagine a 26-D shape... surprisingly we are able to model them with our computers that are less disabled when not able to handle them in their "mind's eye." Our computer models have been able to discern far too many distinct shapes for the extra dimensions, throwing some question as to the capability of superstring theory being the complete theory of quantum gravity.

These extra dimensions of extraordinarily complex shapes exist at the fabric of the universe and are therefore so small that we cannot distinguish their dimensions for most domains of experience. For the domain of experience of you whipping along in a train, the power lines are 1-D, their other two dimensions are wrapped up so small that until we investigate them further, our experience of power lines is 1-D, while the same power line's full, small-diameter cylinder, 3-D nature is the truth of existence. Until we get down to the quantum realm, our existence always has three spatial dimensions, even if our experience "writes off" some of the other dimensions, or more rigorously, we mathematically explain a geometric realm that can only exist in virtual reality, a world where more precise spatial tools can be brought to bear.

Fig. 4 - An example Calabi-Yau Manifold, a forerunner for the 26-D shapes of the extra dimensions on which the superstrings "strum" at the fabric of existence. Note: no matter the complexity of the virtual reality of this shape, you and I can only perceive it in 3-D

Take for example a sheet of paper, it is only the approximation of a plane, not a true plane, it has depth. When we deposit an atomic layer of oxide in a diffusion furnace in the semiconductor processing industry, it is still a 3-D structure, one solitary, molecular layer deep. A wire line made of a single carbon nanotubes is still a cylinder, and even with its perimeter unwrapped, the flattened wall stays three dimensional, one atomic layer in thickness. The existence of material is always 3-D, at least down to where we start approximating quantum particles as *points without dimension* or even further "down" where we begin to look at the nature of the fabric of reality, which we will do in later chapters.

However, our spatial perspective can match geometric reality of different spatial dimensions below 3-D in virtual reality. We will use VR often in this book to model how our experience can be manifested to the computational nature of existence. In the case of perfect, lower dimensional objects, David Deutsch guides us that, "Virtual-reality calipers would have to come to a perfect knife-edge so that they could measure a zero thickness accurately... We *can* perceive perfect circles in physical reality (i.e. virtual reality); but we shall never perceive them in the domain of Forms, for, in so far as such a domain can be said to exist, we have no perceptions of it at all."[25]

Our experience of space can be manipulated in many ways in virtual reality. If you are afraid of heights, your fear of the virtual heights on "Richie's Plank Experience," a virtual reality where you walk out between buildings on a narrow walkway, will be terrifying in correlation with the fidelity of the program to visual and auditory cues (and someday other environmental stimulus like the touch of a wind gust). Advances in technology will likely allow us to develop the weightlessness of space and quantum coding of virtual-reality generators. This claim builds on Deutsch's argument that due to the comprehensibility of existence, we can create a virtual reality generator whose repertoire includes every physically possible environment, even those with higher spatial dimensions, but our ability to experience any physical realities we can code will be limited to our conscious modeling of spatial dimensions, virtual or otherwise. Even in the most advanced quantum VR, we will not be able to experience greater than 3-D space. It is easy to see our limitations when we could code "shape = ," which would compile a 4-D hypersphere in our VR generator, which we would never be able to perceive. Unless we evolve the ability to perceive greater than 3-D space, our ability to know existence's higher spatial order will be limited to the consequences of entertaining higher dimensional space into our 3-D world through our mathematical models.

Like our heroes A. Square of Flatland and Mary of the Black and White Box, our perception of the full spatial realm of existence is at once cognitively knowable and experientially out of touch. This is not surprising given the cosmically brief time and utter lack of selection pressure on our forebears' visual acuity in 4-D, but is it meaningful? Is there something out there in the spatial realms that we cannot see? Is that where we might find Atlantis, the Platonic Forms, or heaven? Unless you like tiny spaces on the order of Planck's length, I would not advise opening a door to the tenth dimension, but I do think there is meaning in just our ordinary three spatial and one temporal dimension of spacetime and it is based in another uncanny trick our mind plays on us, separating us from existence.

Oneness with Spacetime

Check out the space you are in right now. You are sitting or standing on some surface. You are surrounded by air. Items in your visual field have common patterns and come complete with labels: chair, wall, window, tree. Your body is distinct from these other items.

But how far down does this distinction between you and even the air "around" you go? Continue to zoom by a magnification power of ten and the distinction disappears very quickly. At just five orders of magnitude of burrowing, a pore in your skin and a droplet of mist share the same diameter, and at just a nanometer, a billionth of a meter, nine orders down, the atoms of carbon that make up most of your skin jostle with atoms of oxygen and nitrogen in the air.

Our tunneling to resolve this demarcation between us and the air becomes far lonelier, as matter is mostly space, until we find ourselves in the electron cloud of carbon, oxygen, or nitrogen. Here we would no longer distinguish particles by their spacing or some boundary layer, but instead, like some sort of atomic Linnaeus, we count the vibrating protons and neutrons at their nucleus. Here is the last distinction and it is truly one without a difference as quantum interactions become prevalent and erase our hope of imagining a journey to this foreign landscape of the atomic epidermis (also a good name for a band).

EXISTENCE AND EXPERIENCE

As much as we hope to pull back, our mission is fraught, and we race into the quantum world. At a yoctometer (10^{-24} meters), we glide past the neutrino, the smallest massive particle—millions of which typically glide right through the earth without interacting with another solid thing or its field. We still have to travel to a billionth of a yoctometer (10^{-33} meters) before we crash land into the fabric of the spacetime continuum.

Instead of shrinking to get a feeling for the makeup of space, imagine simply walking. On level ground at a moderate pace and with a slightly chilly breeze, you can feel how the air parts around you. At what level does this distinction become fuzzy enough that the material you think of as your body is no longer separating other material and their fields but instead becomes integrated? Forces between most particles that we think of as part of "us" will repel, but if the space we occupy at the start of our walk is an unimaginably small grid, a billion times smaller than the unreactive neutrino, the new space we occupy throughout our walk is not pushed around us but is integrated within us, within the air we displace, and within the solid walls that we cannot pass through. Space is not something we walk on or hover above. We take up space and travel in some way *through* it. At the level of space, all things are one.

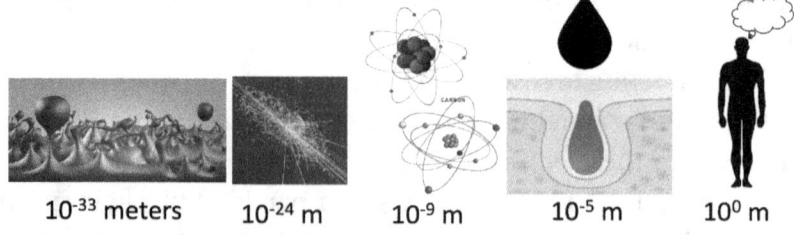

10^{-33} meters 10^{-24} m 10^{-9} m 10^{-5} m 10^{0} m

Fig. 5 - The Orders of Magnitude of Material and Spacetime

For each of the metaphysical theories detailed in the following chapters, the fabric of spacetime is not taken as a simple grid but instead the roiling network of energy strings, foamy relations, or bits of information, but for now the important concept is that the space we occupy is integrated and—at least at the level of the spacetime continuum—there is a unity of all things. Oneness.

Oneness is amongst the most meaningful components of a contemplative's experience. A feeling of a oneness with nature, the universe, or a god is a profound experience. It is a trick of the model of our "selfness" to consider our bodies as solids, separating everything around us and leaving a tumultuous wake behind us. Instead of the experience of solids and space, ourselves and the invisible atmosphere in the room around us, at the smallest levels of existence, it is all space. Spacetime—the special fabric that wrinkles to keep light's speed constant and that warps under the weight of massive objects—stitches everything together. The items in the space around you occupy a spacetime grid integrated with you, our experience of the separateness of space is, at the level of the fabric, an illusion. Seeing yourself as integrated with the rest of world around you, as not being different from the myriad things at the most core, fundamental level, will create a sense of being connected by a context that is not only a background for the activities of your life but that is also the alphabet used for writing the story of existence.[26]

We learn that Australia is not really on the bottom of the world, that velocities, motion, and position are taken relative to the position of the observer, and can finally draw the conclusion that space has the property of non-locality—the x-y-z coordinate grid that we count on to arrive on the correct floor and the correct hotel room, is only part of our macroscopic worldview and is not inherent to spacetime. With our topsy-turvy relationship to "which way is up," we are still part of a continuum, woven together, contained within, and containers for, space.

Fig. 6 - Moving Through the Continuum. The Continuum Moving Through You. Note: The spacetime that the tree occupies now is the same as Mary occupied before the earth moved through it.

Introspection about the way we experience space and the existence of the spacetime continuum is holding up the first of many signposts where meaning is manifest from the multiverse. Our conscious, lights-on experience is unique, a world unto itself, and a known source of meaningfulness, but it can be "leveled-up" when imbibed with and informed by the scientific laws that operate existence. Meditating on the oneness of a continuum necessary in its role in wrinkling to maintaining constancy (the speed of light), warping inward to bring things together (gravity) or "outward" to push them away (dark energy), and as a context for the infinitesimal progenitors (superstrings or quantum foam or bits of information) of everything in existence, it is normal to be awestruck and to feel, once again, insignificant in the face of such magnificence. We will continue to confront this sodden solipsism that gives too much credence to the material nature of existence. In the frame of existence that compares our mass or the space we take up against the rest of the known universe, we will indeed appear to be a minor collection of carbon, but as we will see, computationally we are on a level playing field with other nodes in this massively parallel universal network, and it is this frame, more than any other, that gives rise to the meaningfulness and profundity that comes from our connectedness to everything else in the continuum[27].

As much as we have to learn from our experience of space, it is our experience of time and its nature that will advance our understanding of the inherent meaning available from the natural world.

Timelessness

> *In a world where time is a sense, like sight or like taste, a sequence of episodes may be quick or may be slow, dim or intense, salty or sweet, causal or without cause, orderly or random, depending on the prior history of the viewer. Philosophers sit and argue whether time really exists outside human perception. Who can say if an event happens fast or slow, causally or without cause, in the past or the future? Who can say if events happen at all? The philosophers sit with half-opened eyes and compare their aesthetics of time.*[28]
>
> Alan Lightman

My favorite book to gift, *Einstein's Dreams* by physicist Alan Lightman, plays with different ways that space and time could be experienced. In one of Einstein's dreams, he asks us to imagine that time was just another spatial dimension, so that movement either propelled you forward into your future, repelled you backwards into your past, or, if you sat still enough, you could remain in the infinite present. Through these vignettes, Lightman gets us to appreciate both our experience of time and time's real nature in existence.

Time is not experienced like a spatial dimension... it flows. No matter how faulty your memory or how exceptional your visualization of the future, *the present moment is the only time-*

stamp where you can experience existence. In some way, the present has already gone through the past you remember and has not yet gone through the future you visualize. Depending on your presence, your experience can include all, some, or none of the contents on offer from existence and the flow of time will seem (experientially) to lengthen or shorten depending on how "into it" you are. The experience of a monotone lecture or magnificent sex is (partially) a difference in our experience of the flow of time. A singular focus on the present is one of the most profound contemplative practices available for insights into the smear between experience and existence. At the level of our conscious experience, we can be reminiscing on the past, seemingly transported by a song or a smell to a different experience, yet for all of this memory's verisimilitude and emotional context rushing forward in time to greet us, we are experiencing a thought overlaid onto the present moment. The past is the past. The future is even more amorphous. Our science can make some relevant predictions, and we can hope for our wildest prognostications to come true, with the probability of any event in the future being partially caused by our planning and actions in the present and partially a matter of chance. However, all of our predictions, prognostications, and probabilities are being hatched in the present state of existence. Whether overlaying memories or plans, actively noticing or mindlessly daydreaming, our sensory and cognitive conscious experience commingles with the existence of the universe in the *here and now,* and never in another space or time.

Fig. 7 - Our Experience of the Flow of Time. Note: c is a gauge of our conscious state, delta s is the entropic state of the universe.

The present, where t=now, is transitory in our experience, and still the subject of intense scientific scrutiny. Even in our most mindful moments, the whole of the present can slip away unnoticed and yet—while conscious—we cannot help being on the spectrum of "being present." Our condition of experience is being present. Focusing on the temporal nature of experiencing existence illuminates crucial insights such as the interference of parallel universes responsible for thawing the frozen flow of time, the illusion of free will, and a path to meaning in the multiverse. Let us start by parsing time into smaller and smaller slices in order to determine if our experience of the present is different or insightful from the existence of the present.

The neuroanatomy, signaling, and chemistry of our brain that constructs our continuous experience of existence does so from the past. Another way to say this is that *the present* moment of the phenomena, the brain processing, and our experience are different. In the same way that optical illusions can contort the brain's model of space or physical objects, it is possible to manipulate the brain's delay to show that conscious action is not controlled by volitional executive function by the "self." Instead, this brain model, known as free will, is a temporal illusion utilized by the brain to support our feeling of agency when we need it the most—when confronted with lions, tiger moms, and bears. Studies[29] have shown that the firings of the neurons in the brain cause what we believe to be our willful action before we will them into being. Simply clocking the time between cognition and action shows that our experience of willfulness was started seconds in advance in our brain, the proximate cause of the actions we believe we are directing is unconscious.

There are other reasons to think that, especially as translated from neuronal firings into our experience, free will is a useful evolutionary illusion, the other side of the coin to the illusion of self. Recall that through even the most basic of first-person introspection one can notice that the appearance or disappearance of a homunculi sitting behind your face—your*self*—is still just a piece of content in the context of consciousness. Consciousness is more fundamental than the self. Understanding the self as an idea, meant to act as an evolutionary nexus of opportunities and actions, and not the main event of experience, helps us to relegate free will into the category of mental model with the self. There is no need for an idea (the "self") to have liberty to act *prior to* another sort of brain firing (the action's neurocorrelate). Both are just systems of qualia and, persistent though they may be, neither the unity of conscious behavior (the self), nor this singular self's volition prior to all other neuronal activity (free will) stand up to introspection or the investigation of FMRI study. I believe that the self evolved as a model for acting willfully, quickly, and with singularity of purpose in a highly social world. Just as our brain models a melded experience of sights, sounds, and feelings, it also models an experience where we act with agency on *our* own accord, rewriting the wrongs of our history, and directing our narrative arc into the future and for posterity. The experiential model of self and free will are vestigial for an examined life.

The examined life, especially meditative practices generally described as mindfulness, are central to illuminating the surface selflessness of experience and calming the evolutionary whirlwind of the thought of the ego. Meditation is the difficult practice of observing ourselves thinking without knowing we are thinking. The more concentrated the practice is on observing only the mind, the less the evidence of either the ability to will the next thought into being or a center from which all experience emanates, instead, all experience is consciousness and its contents.

Moreover, not only is our experience out of sync with a self acting with libertarian free will, one in which you could have chosen to do differently if you could go back in time, but even traveling in the plain-Jane forward direction of time, our experience of the flow of time does not jive with the freezing together of time with space into the spacetime continuum.

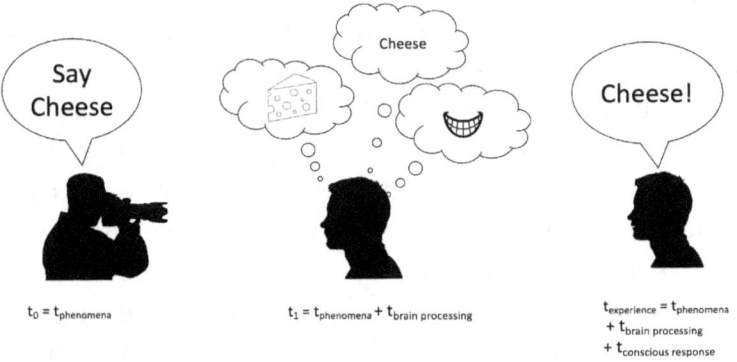

Fig. 8 - Dissecting the present at both the level of the brain and our experience of existence.

The existence of time in *our universe* is more like the spatial time of Lightman's *Einstein's Dreams* than the flow of the present through time. Time in *our universe* is stuck together with space in the spacetime continuum. Any advances that the observable universe makes on whatever it advances upon at the edges of the visible universe is done as conjoined spacetime in what is called the frozen or block universe. In *our universe,* there is no more a flow in time, than there is a flow in the z-direction, where up makes you feel "high" and down makes you feel "low." We know that both the space and time in spacetime are wrinkling to keep constant the numerator and the denominator, position over time, in the equation for the velocity of light; but even as we put a governor on our speed when we launch supersonic spacecraft or even as we move at more human pace, our clocks change relative to a *stationary* observer, although to an almost unappreciable amount in the case of pedestrian traffic. The merger of space and time into the frozen block universe would be indistinguishable from Lightman's account of Einstein's dream except for the thermodynamic arrow of time directed from our low entropy origin at spacetime's beginning—the big bang.

As familiar as we are to the experience of the flow of time, existence also gives a direction to time. The one-way *arrow of time* is tied to the second law of thermodynamics, the law that states that randomness is always increasing in the universe as a whole. This is the most unshakable of physical principles, it is our universe continuing to go on ringing long after the initial bang of the big bang, continuing its progress from the extremes of order of the singularity where everything was smaller than a superstring, to where everything will be separate in the extreme and the universe suffers a heat death. We can use the general entropy of the universe as a calendar for existence, easily able to determine if an event is in our past or our future by how disordered the universe—on average—appears during that event.

Meaning in the Multiverse

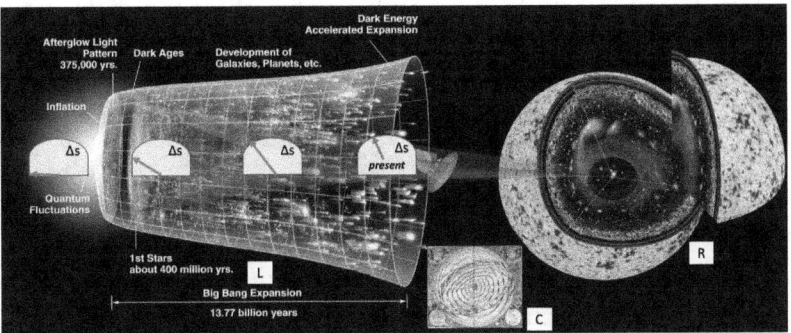

Fig. 9. (R) - The shows a dissection of the visible universe, that when viewed from the present, sketches out a light cone *(L)* back to the point of the big bang singularity (entropy = 0). The expansion of spacetime and the increasing entropy of the universe results from our low entropic start at the big bang and is known as the thermodynamic arrow of time. *(C)* Shows the similarity between our view of the universe expanding away from itself in all directions and the religious conception of the geocentric view of the universe.

We've already said spacetime does not flow, it is frozen; but we also know that causes precede effects in accordance with entropy and energy and chance. The progress of time uncovers a paradox at the heart of both experience and existence: how can something frozen flow along the arrow of time?

In order to begin to defrost the universe and make it more viscous, let's consider our historic universal light cone in Figure 9(L) above as a jigsaw puzzle. We are at a quandary in putting together our new puzzle since we understand where the most entropic bits align, they are determined to fit only in a certain place, given their general entropic nature, but cannot put them down together in *any* order, for that would suggest a flow of time in the creation of our jigsaw puzzle universe. Like a jigsaw puzzle, no individual spacetime slice or puzzle piece caused one another, but, their individual shapes, their entropic character, determine their interaction. According to David Deutsch, "therefore we know that even though some events can be predicted from others no event in spacetime caused another... what we are seeing is that spacetime is incompatible with the existence of cause and effect... it is not that people are mistaken when they say that certain events are causes and effects of one another, it is just that that intuition is incompatible with the laws of space-time physics... but that is alright, because spacetime physics is false."[30] What Deutsch proposes instead is that time requires a quantum conception, one where the puzzle pieces flow together across a stack of parallel surfaces and as such: *The flow of time is actually interference across numerous, more entropic, near-parallel universes in the multiverse.*

The multiverse we are referring to, in all cases in this book, is what is known as the quantum multiverse and stems from the many worlds interpretation of quantum mechanics first suggested by Hugh Everett III. As we will discuss in chapter six, the many worlds interpretation is actually one of the simplest explanations for *how* the well-described, empirically-accurate processes of quantum mechanics resolve to the macroscopic world. The multiverse is a continuum of *near-parallel universes* and *far-parallel universes* to our "own." Near-parallel universes are "near" one another in that there are a limited amount of quantum differences between these universes and this congruency enables them to continue to create interference between them. David Deutsch uses the economic term *fungibility* to describe this property of near-parallel universes. Whereas you might have deposited some of the money in your checking account electronically, in five-dollar bills, or converted them from Yen after a trip to Japan, you do not need to keep track of the individual dollars in your account, they are interchangeable—fungible with one another. Near-parallel universes are the same way, they can have some different properties—like two different dollar bills have different serial numbers, wrinkles, or tears—but the same purchasing power. Like dollars taken out of rotation

Meaning in the Multiverse

by the Treasury and held only by collectors, parallel universes can lose fungibility too. This occurs as the continuous process of quantum distinctions (in accordance with the Schrodinger equation) change the universe to be too dissimilar from others in its previous near-parallel group and, hence, becomes far-parallel to that group. There are an infinite number of far-parallel universes in the quantum multiverse and either a large finite or countable infinity of near-parallel universes. The ideas of both the fungibility and differing properties of near-parallel universes in the multiverse is all that we need to continue our discussion of the flow of time actually being an interference between near-parallel universes in the direction of those fungible universes with slightly greater entropy.

We tunnel through a stack of parallel-universes, disturbed by quantum events that distinguish them from one another[31]. The slices of time we call moments are actually full frozen universes in the multiverse, traversed in accordance to the laws of physics. Before we had the concepts of spacetime, quantum mechanics, or the thermodynamic arrow, our intuitions were tricked in the temporal illusion of the flow of time. We have continued down this path of trusting our under-evolved intuitions even when our classical theories of thermodynamics and relativity failed to create a cohesive explanation for their collaboration. If we really try to interrogate our intuitions, the flow of time is as unintelligible as the flow of the direction "up." While traveling through the multiverse seems strange, it should again not surprise us that a quantum computational parallel-processing multiverse should function in ways counter to our intuitions—what should surprise us is that speaking apes would be able to figure it all out by just observing existence and thinking on it. Stepping back, a multiverse makes sense; our flow through it clears up existence's issue of the frozen arrow of time and our experience of time's flow. It is a big bite to chew and we will be working it down through the rest of the book.

EXISTENCE AND EXPERIENCE

The dynamism of existence and experience depend on *interference* in the multiverse. Which near-parallel universe experience *collapses* in next appears probabilistic, in accordance with quantum mechanics and chance and hence generally follows the thermodynamic arrow of time. There are many possibilities on this continuum, there are most likely paths and incapable paths (those opposed to the laws of physics), causes can be set like traps to define likely effects in the *future*, the path already traversed is remembered, and the here and now—the *only* place and time where we can experience existence—is a frozen spacetime sliver. The length of a moment, the time we are in our particular universal sliver, only extends until the wavefunction changes anywhere in a near-parallel universe and interferes with our own, and then the whole multiverse bifurcates with this distinction. This is a very short time. We are drawn to the present not because it is timeless, but instead because it is fleeting, a place of respite in an ever-changing process. Throughout our lives, we will only ever experience the present sliver of the universe we are in, but we will in some way traverse a near infinite array of near-parallel universes, in human terms almost imperceptibly similar, generally flowing toward greater disorder. The likelihood that an effect will stem from a cause is based on the distribution of near-parallel universes that have said effect in them—increasing the likelihood that our next collapse will be to one of them.

We have always experienced the multiverse; we have just called it the flow of time. We reference previous universes we've visited as the past and plot activities to increase the probabilistic likelihood that we will arrive in universes to our liking in the hoped-for future. Given that the multiverse is subtly experienced as the flow of time, its component near-parallel universes are ever interfering like when old microwaves would increase the static on the radio.

Meaning in the Multiverse

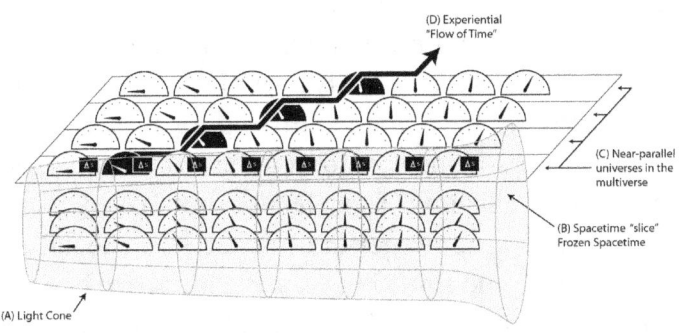

Fig. 10. (A) - Represents the same light cone from Fig. 9 above, with the singular frozen slice of the continuum of spacetime shown at (B). The multiverse is on the x-y plane with fungible near-parallel universes at (C). The only way that we can experience the flow of time in a frozen spacetime is by *tunneling* to near-parallel, more entropic universes in the multiverse, as shown in (D).

We have been thinking about all-natural universal meaning all wrong. Our metaphysics are stuck with a planetary model of the atom, a clunky, billiard ball celestial landscape, and a Copenhagen interpretation of consciousness where we throw our hands up and claim no explanation can exist. Instead, what we will continue to find is a multiverse that frustrates even our most profound intuitions with more wonderment than could ever be created by some desert dime store novelist and natural places to include meaning that neither manipulate existence through pseudoscience nor inundate humanity with a specialness in the cosmos we do not deserve. Like the flow of time really being a traversing of near-parallel universes, what we thought of as solely personal meaning—experiences of flow, mindfulness, or other sorts of profundity and optimized well-being—are processes run in existence, optimization programs on a massively parallel quantum computer we call the multiverse. As we have investigated where experience and existence blend together as context and contents; warp, wrinkle, and weave together all things into a single continuum; and bifurcate in what we experience as the next moment, but what is actually a whole different (more disordered) universe, we have opened up opportunities to let all-natural meaning—for so long considered the solely held property of experience—into existence as well. Not only have we clobbered our intuitions about our experience of and the true nature of the existence of space and time, but we have put them back together more whole, in a way that fits the latest physics, and in a way where we can better understand the challenges this untrodden path is taking us on. We have forced the door open using speculations on theoretical science and philosophies of the mind that in their own right still have staunch opponents and will still go further to explore these speculations and their consequences. This is a rabbit hole down which academics cannot go. Careers in science and philosophy must closely guard their credibility and this must be respected. Humanity requires science to adhere to protocols so it can open our eyes when the right explanation and clever experimentation come together, and not a moment before. However, the same respect must be given to the speculative, for I'm not saying with certainty that this is the way that things are, no one can do that, but what I am stating is that it is reasonable to revisit our current scientific conceptions of reality and question if they contain meaning. If we are going to ask the question, "what is the meaning of it all" anyway... it is best to leverage the latest science of existence and the latest interrogations of experience.

Updating our metaphysics, the philosophy of existence and experience, is the first step. Thankfully, our metaphysical hypothesis of the fundamental components of the natural world are not so much wrong as outdated; material, idea, and process ontologies are less combative than collaborative, build into a crescendo that has the possibility of arriving at a simple statement of what existence is, how it programs our experience of it, and how it is meaningful. We start with the dominant metaphysics of our time, materialism, the idea that some form of stuff, matter, is fundamental in our universe and ask how a materialistic universe might impact universal meaning. We follow the same line of conceptual introduction and its meaningfulness when considering if information, not material, is fundamental. Finally, we introduce the concept by which material and information are emergent from the laws of physics in a computational universe where meaningfulness is an optimization routine run across universes in the multiverse. Thinking differently about the makeup of the world can change our minds about whether meaning can be found "out there" and lets us form a reasonable hypothesis about existence's role in optimized experience.

CHAPTER GLOSSARY
Existence - the physical world as defined by the laws of physics.

Experience - our first-person conscious experience of sensory inputs and thoughts.

Many worlds interpretation of quantum mechanics or the multiverse - Hugh Everett III's solution to the measurement problem that explains the collapse of a particle from all the possibilities of the wavefunction as just a parochial view of wavefunction evolution across many parallel universes that collectively represent all states of the wavefunction.

EXISTENCE AND EXPERIENCE

Hard problem of consciousness - subjective states do not appear to be constructed from the "bottom-up" or from some more fundamental informational processing mechanisms.

Thermodynamic arrow of time - the future will be more disordered than the past, following the propulsion of entropy increase that has been the fundamental fact of the known universe since its beginning at the big bang.

KEY TAKEAWAYS

- Investigation of experience allows us to deduce that the *context of consciousness* is the only thing which we cannot be deceived about; however, the *contents of consciousness*, including our perception of the existence, can be counter to our intuitions.
- Consciousness is constructed from our felt experience and our inner narrative of ourselves. While this inner narrative and our thoughts, broadly known as *the self*, has been essential to our developing a rich society, its overindulgence can lead to unsatisfactory mindless states that react erratically.
- Investigation of *existence* is what we call science. Our intuitions of existence can be flawed and only a strict adherence to designed experiment, reason in analysis, and repeatable conclusions allow us to form an understanding of the objective facts of existence.
- Existence and experience have a few paradoxical relationships whose inquiries are fundamental to a more accurate model of the world and to the nature of meaning. Our experience of the locality of space and the flow of time is countered by the more meaningful fundamental reality that space connects everything and that we "flow" not through "time" but instead experience interactions between near-parallel universes.

PART II:

Metaphysical Hypothesis

CHAPTER 3:

Materialism

CONCEPTS
- Materialism, the world's most widely accepted model of existence is introduced. Materialism's greatest empirical efforts like quantum physics and relativity are explored.
- The vanguard of materialism—quantum gravity—a unification of the fundamental forces and a complete particle guide is introduced.
- Are dark matter and dark energy material? Is materialism capable of detailing a periodic table of force-carrying and massive dark particles? We will look into what dark matter and dark energy suggest of the multiverse and repellent energy from nothingness.
- The personal nature of mindful materialism is discussed. This meaning is given greater significance and is universalized by considering the rarity of conscious experience in the multiverse.

THE SOMETHING, OR THE stuff that exists, that we are most intimately familiar with, is matter. If someone tells you there is no god, no universal consciousness, no mind separate from the brain, or (heaven forbid) no need for metaphysics, that person is likely a materialist. Materialists have reason to be cocky for it has been materialism that has dominated the major findings in science over the last century.

Materialism had auspicious beginnings. The fundamental components of the world were thought to be earth, wind, fire, aether, and water—the classical elements of Ancient Greece. Other cultures maintained a similar set of fundamental elements and worked them into similar tables of comparison. An element could be generative with respect to one or more of the other elements, but corruptive with respect to others; their reactions could be subtle or intense.

This table of elements was put into practice in alchemy. More miss than hit, alchemy tried to manipulate the properties of materials to make more desirable materials without the benefit of a correct underpinning of the fundamental mechanisms, without well-designed experimental procedure, and with secrecy that kept the unlikely successes within a small lineage from tutor to student.

Since Max Planck formalized atomic properties with early quantum mechanics in 1900, the quickly developing picture of more and more fundamental constituents of stuff, their relations to one another, fields they create, and their causal linkages has been responsible for the increasing complexity and technological sophistication of the modern age. Not only have physics and chemistry largely been built on materialistic underpinnings, but so too have medicine, genetics, and neuroscience benefited from a fundamental periodic table approach.

Materialistic metaphysics claims that both mental and physical processes are the result of matter's interactions. Thomas Hobbes, a contemporary of Descartes, argued that mental events are caused by motions both inside and outside of our brain, three centuries before neural plasticity would win a Nobel Prize.

As I sit and type, my fingertips hitting on the keyboard are the part of me that interacts with stuff. Inside my computer, when I hit "command + S," electrons (a very small sort of stuff) quantum tunnel across a few molecular layers of silicon dioxide ("wafer rust") to rest on an accommodating, conducting cell of metal surrounded by an insulator. The many electrons sitting in many cells short circuit a channel[32], indicating the storage of memory—my book. My eyes interpret the changing landscape of my screen, sending electrons racing along my neural pathways. The energy of electrons is exchanged inside an organic lattice of nerve cells until it reaches the end of a dendrite where electrical potential is exchanged for chemical potential, which disperses into the synaptic gap between one brain cell and another.

> *If real is what you can feel, smell, taste and see, then 'real' is simply electrical signals interpreted by your brain.*[33]
> Morpheus, *The Matrix*

Neuroscientists are demolishing Cartesian barriers between material interactions in the brain and the mental processes normally attributed to *the mind*. The idea that the brain is the seat of memory, emotion, sensory response, and logical processing is the simplest answer and is backed up by the work of psychologists, physiologists, and neuroscientists. Developments in the treatment of mental conditions like depression and schizophrenia using neurological drugs further suggests a linkage between the brain's biochemical processes and mental states.

Strict materialists, while not yet able to completely describe consciousness from brain functions, feel that our understanding of computer operation can help explain mental processes. They believe that the interaction between the physical components of transistor gates, capacitors, and circuitry and the programs that logically operate them is analogous to how our own machinery works. In Daniel Dennett's book *Consciousness Explained*, Dennett claims that our understanding of how the brain functions dissolves the consideration of the mind as different from the brain computer. "The brain acts as an organic machine that uses effects of cultural inheritance to store and utilize information and to solve problems."[34] Most materialists feel that brain biochemistry has adequate processing power to fully explain all of our conscious states.

Material existence is extraordinarily well-detailed. Modern neuroscience is rapidly developing a clear understanding of both deficiency and acuity; while physics has detailed the fundamentals of matter and its interactions while giving clear predictions about the original makeup of the universe. The general theory of relativity and the standard model of quantum mechanics represent humanity's most accurate understanding of existence.

When stuff shimmers: Wave-Particle Duality

Human investigation of the material world, the world of solid bodies, was well-formed by the start of the twentieth century. Galileo, Kepler, and Copernicus arranged and rearranged the heavenly bodies and Newton showed how their motions and the motions of earthly objects could be mathematically derived. The material world sat against a backdrop of a constant space and time. What you could see could be calculated to great precision with the newly acquired calculus. But there were an increasing number of microscopic cracks in this facade: the body was starting to be broken down into cells; fields that subjected the perceived world to forces were being discovered and engineered to power the world; and there was a periodic nature to the elements that suggested a deeper structure.

At the turn of the twentieth century, physicists were interested in the radiation from a specific type of object, a blackbody, that absorbs all radiant energy and that emits at all frequencies with a spectral energy distribution dependent on its absolute temperature. A clue—in the form of the broken bars of discrete energies that form the spectrum of light from a blackbody—was delivered to theoreticians to derive an equation for this fundamental process. At first, there were failures on both ends of the spectrum. Wilhelm Wien derived equations that matched the experimental data only for short wavelengths. At around the same time, one of the most revered of nineteenth century physicists, Lord Rayleigh, derived an equation that correlated at long wavelengths but resulted in infinities at short wavelengths, a result deemed the *ultraviolet catastrophe*.

Theorizing that the vibration of atoms was giving off energy in the form of radiation, Max Planck discovered that the modes of the vibrations and their frequencies can only take on certain values. Instead of the continuous losses in potential energy like when a rolling stone descends a mountain, think of the discrete losses when the same stone falls down a flight of steps. The full impact of Planck's discovery would not be brought to light for five more years when in an experiment studying the photoelectric effect, another form of radiation under study was also quantized.

In the photoelectric effect, the ejection of electrons from a screen under fire from a beam of light is counterintuitive when modeling light as a wave. Electrons are not eroded away by larger intensity or brighter light, but instead by the distance between the waves, their wavelength. At the level of the screen, if light only acts as a wave, this would imply that it is not the increasing height or energy of the wave that increases erosion but just the speed at which light waves break against the screen. Einstein's proposal was that all light was made up of energy quanta, photons, which, when impacting the screen in discrete packets, ejected more electrons as more of them hit the screen, like particles fired from a light gun.

Max Planck broke from convention to solve the blackbody radiation equation by quantizing the wave-nature of radiation; Einstein explained the photoelectric effect by quantizing the particle-nature of light, but this was still not a leap far enough for Prince Louis de Broglie of France. In 1924, he submitted a theory that matter should also be considered as both a wave and a particle. Things started happening very fast after this discovery as the fundamental nature of both the energy released from matter and the structure of the atom itself was unmoored from what we could see and experience. Indeed, the modern understanding of the nature of the material world is almost wholly unrelated to our experience of it.

That radiation, light, and matter act as both a wave and a particle has been proven in experiment and is also well-codified in the mathematics. Even as the discovery of the wave-particle duality of light and matter piques our curiosity, the inner-workings of the quantum world are far more surreal than could be imagined by even Salvador Dali. The nature of existence is far different than the model we have of it in our mind.

Quantum Smallness

For most of us, the picture that comes to mind when we think about the most fundamental components of matter is the atom. According to quantum mechanics, the atom is not a nucleus planetoid orbited by distinct electron moons, but instead a balloon-animal-shaped cloud of electromagnetic force. Quantum electrodynamics explains the interaction of the electron with the force carrying particle of electromagnetism—the photon. The photon is a fundamental particle, classified as a gauge boson, a force-carrying particle. The exchange of photons between electrons, a matter particle, begged the question of whether other fundamental forces could be explained by the exchange of quantized components of virtual particles.

In the 1960s, quantum field theory gained its first verified combination of fundamental forces when electromagnetic and weak nuclear forces were shown to be part of the same force named the electroweak force. Similar to how photons are exchanged by electrons to generate electromagnetic force, weak nuclear force is the exchange between the matter particles (the neutron and proton) of their force carrying particles the W and Z bosons.

Next, the strong nuclear force was studied and its interactions—this time between the matter particle triplets of quarks and the force carrying gluons—were found to fit well within quantum field theory. The strong nuclear force is explained by a branch of quantum field theory known as quantum chromodynamics because of the more complex way that quarks must be grouped together (into "colors") to form a stable particle or a hadron. A hadron must be "colorless"—an arbitrary way of discussing a property of a quark triplet that contains each a red, blue, and green quark. For example, the hadron known as a proton contains two up-quarks and one down-quark and each is a different "color." At the Large Hadron collider at CERN[35] in Switzerland, distinct quarks can be detected momentarily before they degrade into hadron jets and electrons.

The standard model of quantum field theory (or the standard model) is a quantum field theory periodic table of sorts, ordering and predicting the various fundamental components and their dynamic interaction in the electromagnetic, weak nuclear, and strong nuclear fundamental forces. The agreement of quantum field theory to experimentation is phenomenally accurate (to 1 part in 100 million),[36] giving materialists claim to the most empirically-verified understanding of the nature of the cosmos.

Meaning in the Multiverse

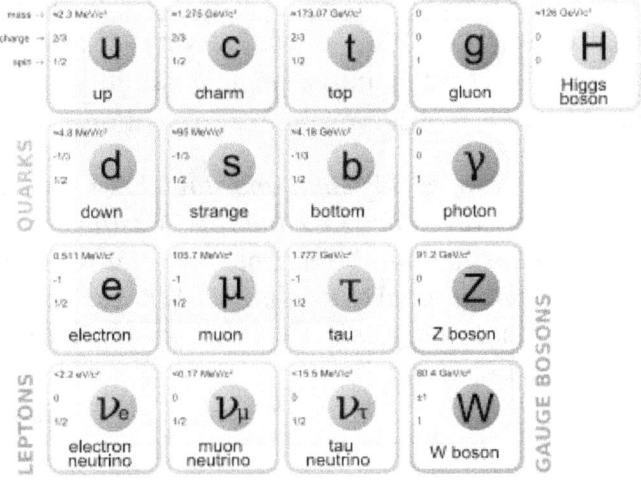

Fig. 11 - The Standard Model

Elementary particle physics is far from standing still and even with the recent (2013) experimental confirmation of the Higgs boson at CERN, the standard model is being utilized as the building block for more exotic hypothesis that "incorporate hypothetical particles, extra dimensions, and elaborate symmetries (such as supersymmetry) in an attempt to explain experimental results at variance with the standard model, such as the existence of dark matter and neutrino oscillations."[37]

Even as the simple planetary model of the atom was fragmented into the more fundamental constituents of leptons and bosons, having more fundamental particles was nothing new... it is how these particles interact and their fundamental nature that draws out some of our most interesting, intuition-expanding, and meaningful explanations of existence.

The way a fundamental quantum particle evolves, as detailed in the Schrödinger equation, is a perfect example of a process of existence that is inconceivable at our level of experience. The Schrödinger equation defines the *wavefunction* or the state of a particle in *superposition* with all of its possible states before it collapses to become an actual particle. The wavefunction describes matter as a probability of states, while what is called the *collapse of the wavefunction* into a single possibility is what gives us the particle we can observe and experiment on.

> *I should come clean at this point and state that no one really knows what the wavefunction actually is. Most physicists regard it as an abstract mathematical entity that can be used to extract information about nature. Others assign it to its own, very strange, independent reality... it is [also] possible for both views to be valid.* [38]
>
> Jim Al-Khalili

Let us consider a simple example of a particular particle's spin. As you can see from the standard model graphic above, spin—the third number in the column to the left—distinguishes quarks and leptons (both of which have half integer spin) from their force carrying gauge bosons (which have full integer spin). These spin values can be either positive or negative. An electron, classified as a lepton, can be fully defined by a wavefunction written as the Schrödinger equation, that equally defines a positive 1/2 spin, a negative 1/2 spin, and a positive and negative 1/2 spin "at the same time" or in *superposition* with one another. Before the electron in this case is interfered with by any of a number of particles or collections of particles (in this universe or a near-parallel one), its property of spin can be said to be in a superposition—positive-negative.

How existence arrives at the distinction in our reality between one of the three states defined in this superposition trinary has driven theoretical physics to develop in turn a strident "shut-up and calculate" position that attempted to avoid explanation, an explanation that puts observers into the driver's seat of wavefunction collapse, and the completely new and expansive ontology of many worlds, each with their own evolving universal wavefunction.

The former "shut-up and calculate" approach to quantum physics was forwarded by some of the most august founders of quantum physics, Bohr, Heisenberg, and Schrödinger, who set rules known as the Copenhagen interpretation, stating that since quantum physics was so powerful and predictive there was no need to delve into interpretation. The sheer formalism of the Copenhagen interpretation dissuaded many theoretical physicists from speculating on what the wavefunction was or how superpositions collapsed. In fact, the most influential explanation of wavefunction superposition—what came to be known as the many-worlds interpretation of quantum mechanics—was buried for years and its creator, Hugh Everett III was discredited enough after the 1956 publication of his paper, "The Theory of the Universal Wave Function," that he left physics for computer science.

That a particle is explained as a probability is intriguing and counterintuitive, that *observations* can set matter's state and determine the collapse of the wavefunction is so problematic to the subject-object relationship of experiment and technician that it is called the measurement problem. In our electron spin example, changes to experimental setup can show that taking a measurement interferes with the state of a quantum system and causes it to collapse. There are a couple of different levels of the measurement problem, the easy measurement problem is the spurious claim that there is something special about human or conscious observers that bring about the collapse of the wavefunction. These types of claims, often built by New Age proprietors of woo for the sake of selling scientific-sounding snake oil, are problematic for metaphysical speculations that attempt to walk the tightrope between skepticism and speculation. The easy measurement problem is a bastardization of the mechanisms of wavefunction superposition collapse, and, contrary to popular woo-ology, human consciousness is not required to enable the world to exist. However, the hard measurement problem does present a difficulty because the Schrödinger equation predicts that we should observe all three quantum states, particle, no-particle, and the superposition of both, the last of which we rarely observe.

One explanation for why we rarely observe the superposition is that the wavefunction of the quantum system becomes entangled with the more numerous and complex wavefunctions of the measurement device. This decoherence also explains why natural quantum processes, like radioactive decay, do not sit in states of superposition waiting to be measured but instead collapse through decoherence with the system (the surrounding air, rock, or other material) that was not deliberately isolated from these processes by savvy experimentalists in lab coats. The macroscopic world and even the most adroit observer will always cause the dilution of entangled wavefunctions known as decoherence, resulting in a measurement of a collapse.

> *Larger things tend to be more susceptible to decoherence than smaller ones, which justifies why physicists can usually get away with regarding quantum mechanics as a theory of the microworld. But in many cases, the information leakage can be slowed or stopped, and then the quantum world reveals itself to us in all its glory.*[39]
>
> Vlatko Vedral

Decoherence does a good job of explaining the physical reasons why the wavefunction of particles appear to collapse during measurement with macro-systems and solves the hard measurement problem but fails to probe wavefunction collapse at atomic levels. The simplest explanation for what is going on during the collapse of a wavefunction from the superposition of all of its states to the state that occurs in our universe is one that assumes no wavefunction collapse, but instead the "interaction[40]" of near-parallel universes running with a wavefunction distinguished only by this single state change. In "our" universe, the wavefunction evolves with a positive spin particle in my lab, in the near-parallel universe, the wavefunction evolves with a negative spin particle in an otherwise complete replica of my lab, myself, and my universe. I can only experience the universe I am evolving with and the interference of near-parallel universes, but the mathematics behind the evolution of the wavefunction is consistent with what has come to be known as the many-worlds interpretation of quantum mechanics (which does not require collapse mechanisms). The many-worlds interpretation states that the wavefunction evolves for the entire universe as predicted by the Schrödinger Equation and any "collapse" into one or another state is a distinction—a bit—available only to those experiencing a single universe. The many-worlds interpretation got rid of one issue and replaced it with another.

We will continue to return to the quantum multiverse or the many-worlds interpretation of quantum mechanics thanks to the unique way it draws together computation and quantum physics, our experience of the flow of time and causality, and any meaning we can draw from existence. However, the strict materialist uses the many-worlds interpretation as a means to remain consistent within the bounds of the wavefunction and considers the parallel universes not as separate nodes in a massively parallel quantum computer but as discrete islands separated somehow from one another by the ongoing and ever-present evolution of the wavefunction of each universe. For the time being, we will take their lead and resume our introductions of the most well-verified scientific theories, working from the highly accurate theories of the small (quantum physics) to the equally successful theories of the cosmic (relativity).

Cosmic Largess

Immense in both space (93 billion light year diameter) and time (13.8 billion years old), our visible universe is comprised of three perceivable spatial and one temporal dimension termed spacetime. The spacetime continuum is often imagined as a sheet pulled taut to demonstrate the action of gravity based on its curvature, but this image is harmful to our perspective of spacetime, for we do not travel around the sun on top of the curvature of spacetime, but, as already discussed, instead in some way *through* the continuum. Furthermore, if we could gain an insight outside our universe, we would see that not only was space stretched in this sheet, but so was time. *Relativity predicts that the cosmos is determined*—a frozen block universe—and that motion in space, if done at high enough speeds or near a gravitational well like a black hole, can change or even stop the passage of time. We'll revisit this perspective a little later.

Relativity has created an entire language, a meme about how nothing is set and all must be described in relationship to the observer. Everyone gets it, the thought-experiments are clear and compelling. When a train travels away, the pitch of its horn, that is, the distance between the sound waves in the medium of the air, becomes lower because the waves are further apart. The opposite is true for the pitch of a train speeding toward us, the waves come at shorter intervals creating a higher pitch. The distribution of sound waves in air is not constant nor relative to the motion of the observer.

However, light does not act in the same way. Its velocity is constant. The light from a distant star and the light produced from the above lightbulb is always, in every instance, traveling at 3×10^8 meters per second.

The clue that Einstein understood is that the only way that light's speed could be constant but still have wavelike properties relative to position was if the other component of the speed equation—*change in position/change in time*—were not absolute, especially at the speed of light. In our everyday life, like with so much of physics, we do not see how position or time can be relative, but experiments in Einstein's time were able to show that clocks on one plane flown against the direction of the Earth's rotation to achieve a lower speed, compared with another flow in the direction of Earth's rotation to achieve a higher speed, came back with slightly different readings. It was the thought experiments, like the idea of a train traveling at the speed of light with an observer on the train and one standing on the platform, that really sunk the idea of absolute time.

Einstein further postulated that not only were both time and space not absolute, but they together comprised the continuum that was the medium that light traveled through. The now famous spacetime continuum beat out the idea of aether for the medium used by light waves. During eclipse observations, it was shown that the light of a far-off star traveled along a curved path around the gravity well of the sun—exactly as was predicted by general relativity.

Unlike the standard model with its probabilities, the theory of relativity explains the fundamental force of gravity using geometry. Spacetime is a landscape distorted by massive objects—the more massive, the greater the curvature. It is not until we begin to look at the very smallest of scales—at the quantum level—that we begin to see that spacetime is also a roiling mass of probability. At the cosmic level, planets, stars, and galaxies distort the *path* of another entity in proportion to their mass and the distance between the two objects, as Newton had calculated. Einstein improved upon the idea of an invisible force by tying gravitational acceleration to the geometry of spacetime.

Further developments in the science of spacetime came about when, like kids, physicists tried to break their new theory. Our understanding of the properties of black holes are a result of the study of the extreme curvature of spacetime. In the event of an implosion of a supermassive star, this large density (the mass of a supermassive star divided by the tiny space the imploded star takes up) warps spacetime, creating a massive gravity well from which nothing can escape, not even light: a black hole. It is inside this black hole—where great masses are combined together with extremely small spaces—that the equations of relativity break down into what is called a singularity.

Black holes are only one place where singularities exist in the equations of relativity—the other is at the instant of the creation of the universe: the big bang. At the big bang, the entire mass of all the material in the universe was compressed into spacetime no bigger than a quantum particle. It is at this point in the history of the cosmos that the forces that now act over quantum distances (strong nuclear, weak nuclear, and electromagnetic) and those that now act on cosmic distances (gravity) worked in cohort. The results of the expansion of spacetime—the development of matter, energy, antimatter, and the plethora of differentiated material both "light" and dark—owes itself to this event, this marriage of the quantum and relativistic.

Materialists have seized the opportunity of the singularity to explain the unification of forces. Something other than point particles are required at the material endpoint of the singularity, setting the stage for the smallest of stuff to be described as strings of energy. For decades, much of the effort of quantum physics plunged into describing these extraordinarily small loops. Superstring theory unifies the "notes" playing on superstrings that spawn the different quantum particles with the claustrophobic relativistic spacetime geometry at the singularity. However, superstring unification theories have not yet had the profound success of quantum electrodynamics and quantum chromodynamics in uniting fundamental forces or predicting standard model particles. These apparent failures are due in large part to the extreme smallness of the building blocks (on the order of Planck's length or 10^{-36} m) and the energies a standard hadron collider would have to achieve to investigate them (one with an outer track fitting comfortably around our entire galaxy!).

Is Dark Matter Still Matter?

Materialists have an extraordinary lead in explaining the nature of the universe. Approaching nature from the largest reaches of the cosmos using relativity, the most elegant geometrical explanation of the nature of both space and time, allows us to explore and predict all the way back to the first light of the early universe, while the spacetime continuum's most extreme curvatures are giving us insights into the nature of the link between information and matter. At the smallest extremes possible, quantum physics has been extraordinarily successful and perplexing at once. Using the standard model, continued successes are being recorded in aligning the fundamental forces of nature, while exotic and elegant solutions are probing the very smallest, very earliest moments of the universe where quantum forces predominated.

However, all of these materialist mechanisms play on a smidgeon of the known mass-energy balance of the universe. The standard model of quantum physics, for all of its formalism and precision, fails to describe the behavior of 95.1% of all the matter-energy in the universe. Dark matter and its more elusive sibling, dark energy, continue to perplex materialist physicists and as of the time of this writing, there is little understanding of the characteristics and fundamental nature of dark matter or dark energy.

Dark matter comprises 26.8% of the total mass-energy in the universe. It is *dark* because it does not interact with electromagnetic energy (like light) that might illuminate it, but still *matter* in that it interacts through gravity and the weak nuclear force. The interactions of dark matter give the universe its heterogeneous mix of galaxies, galactic clusters, and empty space. Adding to a further sense of awe in the possibilities of the universe, astronomers have recently found a galaxy, Dragonfly 44, with the same mass as the Milky Way Galaxy, but only 1% the radiation, suggesting that Dragonfly 44 is almost entirely made of dark matter.

The impact of dark matter can be observed in the movements and makeup of the cosmos. Dark matter appears to create a halo around galaxies as observed by the difference in orbital velocity of the outermost stars of galaxy versus those predicted. The halo of dark matter around the Milky Way keeps planets like ours at the extreme edge rotating at the same speed as those at the center, as opposed to at a much slower rate as predicted without the halo. According to astrophysicist Katie Freese, "the strongest, most direct evidence for dark matter comes from the collisions of clusters. Not only is the gravitational effect misaligned from the normal matter, but its magnitude is some five to six times greater than the normal matter would lead us to believe."[41]

After Hubble's astronomy found that the universe was not just expanding but racing away from us, dark energy was used to explain this effect. Unlike matter-on-matter or even matter-on-dark matter interactions, dark energy acts repulsively. Much like inflation, dark energy is used to explain expansion of spacetime in the presence of the attractive force of gravity.

The nature of dark energy is likely the most active field of theoretical physics today. A theory of the mechanisms and fundamental constituents of dark energy would explain the acceleration of the universe and potentially its ultimate ends. Could the universe continue to expand forever, spreading all the matter out along with it? Could there be a contraction and *big crunch?* Is spacetime wrinkling, ripping, or warping back on itself?

The leading theories for dark energy are that it is just the energy of empty space, energy created from vacuum fluctuations as predicted in numerous theories of particle physics. This energy is not in violation of the first law of thermodynamics, which states that energy cannot be created or destroyed, as the energy arises from the same quantum fluctuations possible in the void of space thanks to a vacuum—meaning the absence of particles *but not* the absence of a superposition of particles and vacuum at the same time.

The *darkness* of dark energy makes experimentation complicated, while its repulsive nature suggests mechanisms that are not limited to just matter but the nature of the fabric of space-time itself. Different theories of dark energy, such as the one in the quote below, suggest that this repulsive force is something more akin to the imperceptible arrival of the fourth spatial dimension into our universe.

> *Intriguingly, the simulations also hint that soon after the Big Bang, the Universe went through an infant phase with only two dimensions—one of space and one of time. This prediction has also been made independently by others attempting to derive equations of quantum gravity, and even some who suggest that the appearance of dark energy is a sign that our Universe is now growing a fourth spatial dimension.*[42]
>
> Zeeya Merali

Like in Edwin Abbott's satirical book *Flatland*, where our hero, A. Square, is visited by a higher-dimensional being known as A. Sphere (whose movement through Flatland consisted of first, a point followed by a series of increasing diameter circles and second, a series of decreasing diameter circles followed by a point), the evolution of the fourth spatial dimension into our cosmos may be met with equal unforeseen consequences.

Materialistic Meaning

Materialists describe the fabric of reality as an ever more elaborate interaction of particles and even strings of energy. Materialistic metaphysics explains the most fundamental physical entities in terms of their symmetries, the breaking of symmetries, and their supersymmetries. Symmetry might be proposed as a universal material meaning. Symmetry suggests that predicting the end state of the universe based on its current state is parochial to the *handedness* of our half of history[43]. We have only done the math for less than half of history and only beyond 181-degrees in a cyclic universe can we know the true nature of the entirety of our universe. Symmetry-switching toward a cosmological *big bounce and the taming of entropy* may be driven by opposing "light" and dark energy symmetry switches, or, as we will explore later, reaching some sort of universal dialectic.

Even taking on the unlikely nature of a symmetry switching event for argument's sake, the universe does not appear to be symmetrical in human domains nor does a desire for symmetry appear to be much of a driver for our purpose other than the wholly aesthetic ones. We simply do not drive to have a positive experience for every negative one—we will take the good with the bad as we are forced to—but seek out the optimal experience on its own terms, not as a way to cancel out some past or future ill.

Materialistic metaphysics is the most austere to the development of universal meaning but offers itself most easily to an awe-struck and even ultimately important personal meaning. Whether you are discovering the nature of consciousness through secular mindfulness meditation, exploring the comprehensible nature of the fabric of reality through physics, or simply staring into the night sky, being amongst such grandeur and being able to be interested is enough to fill lifetimes with meaning. As Sam Harris states, "failure to find meaning is largely just not paying attention" and it is true that everywhere we look, including paying close attention to something as (hopefully) ubiquitous as the breath, we find wonderment, mystery, and deep and reliable personal meaning. In our first few chapters together, we have traveled to the extremes of the known universe, learned that we may have near-parallel neighbors that are extraordinary replicas of ourselves, and broached the mystery of why it is at all *like something* to be us, all within the confines of our thoughts. As you will continue to see, my speculations about the universe are of a substantially lower quantity and quality than my appreciation for existence, its comprehensibility, and its relationship to experience—peak experiences and otherwise.

Materialists are not above speculation that the universe cares what humans experience! Indeed, the idea that, to quote the Grateful Dead, "you are the eyes of the world," is not an uncommon way that skeptics add great significance to our meager existence. The idea is that, for all of its grandeur and precision, the only way for the universe to experience and take stock of itself is through the rare and chance creation of sentience in far flung regions of itself. Our consciousness and the stewardship of it through some number of epochs of the universe is *the most important thing* in the entire universe, it is the meaning *for* the multiverse. This form of poetic naturalism[44] is an important part of the story of meaning and personal profundity. The importance of opening some space to just experience goes beyond the benefits of mindfulness and into the universe living vicariously through us. Our personal experience of existence, compiled in any way that it is compiled, amounts to the "greatest story ever told." Nothing else has the capability to build the poetic out of the natural; as far as we know as strict materialists, it is all naturalism.

MATERIALISM

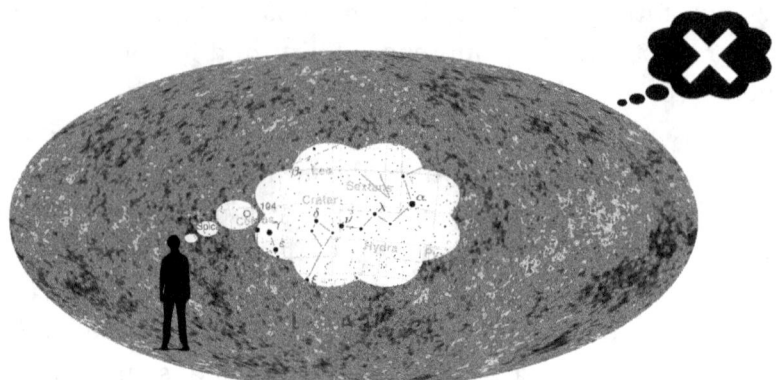

Fig. 12 - **The Eyes of the World Meaning**. Entities with experience offer existence a unique story of itself. Humans are one sort of entity that, through our appreciation, awe, and contemplation of the world, give it meaning and are therefore ourselves very meaningful to the universe.

Most materialists won't go so far as to state that there is anything that the universe can do to nurture this gem of felt experience in the minor characters living in far flung regions of its setting, but that our subjective information processing is given such primacy amongst all of the universe is already an important hybrid of universal and personal meaning. Even though there is something materially happening in our brain that is either wholly or partially responsible for our advanced—poetic—brand of consciousness, as the hard problem of consciousness states, the interaction of those quanta, atoms, and molecules have no business producing experience, it is a deep mystery—one the universe is counting on if it is not already supremely conscious itself. We investigate whether there is anything to a fundamentally conscious or informational universe in the next chapter.

CHAPTER GLOSSARY

Materialism - the most scientifically reliable ontology that has shown that the fundamental nature of existence is material stuff and the fields their interactions form, and that the neurophysiology of the brain is responsible for experience.

Quantum - Having a distinct quantity, as opposed to continuous. The energy of force carrying particles is discrete or quantized.

Wavefunction - Defined by the Schrödinger equation, the wavefunction of a quantum particle completely describes its state "before" it is observed and is the most fundamental equation in quantum physics.

Measurement problem - The difficulty in explaining the evolution of a wavefunction from the superposition of states to the "collapsed," distinct particle state that led to a schism between formalism and explanation in quantum mechanics.

The standard model of quantum field theory - the explanation that details the interaction of all quantum particles, their fields, and three of the four fundamental forces of nature (strong and weak nuclear and electromagnetic).

The general theory of relativity - a geometric theory of gravitation that describes how spacetime is curved by massive objects and how, in order to maintain a constant velocity for light, space and time warp for anyone relative to the travel of light.

Supersymmetrical string theory - also known as superstring theory is a hypothesis of quantum gravity that postulates that all quantum particles arise from the vibrations of tiny strands of energy (strings) fit over complex-shaped, supersymmetrical extra-dimensions.

Dark matter - comprises 26.8% of the total mass-energy in the universe. It is dark because it does not interact with electromagnetic energy (like light) that might illuminate it, but still matter in that it interacts through gravity and the weak nuclear force.

Dark energy - a repulsive force that accounts for 78.1% of all matter-energy in the universe and is responsible for the repulsive acceleration of the universe. It is hypothesized that dark energy is not energy at all but the expansion of a new spatial dimension into our 3-D world.

KEY TAKEAWAYS
- Materialistic metaphysics has been the most important contributor to the science of the natural world and the technological revolution. Most of what we understand about the universe and its evolution is based in what we understand about a handful of quantum particles and their interactions.
- Spacetime is central to materialistic existence, it is shaped by matter causing cosmic gravity, it warps in such a way that light's velocity is universally constant, and, according to string theory, at each stitch in its fabric it roils with loops of energy played by complex extra-dimensions.
- Materialism has nearly unified the fundamental forces into one theory. Quantum chromodynamics and the standard model has unified electromagnetism, weak and strong nuclear forces. A theory of quantum gravity, a theory that would explain singularities (like those in black holes and at the start of the big bang) where matter impacted by gravity is constrained to quantum sizes, remains elusive.
- Materialism's successes only explain about 5% of *material* in the universe. The majority of the manipulation of

spacetime is done by dark matter and dark energy which are difficult to investigate empirically.

CHAPTER 4:

Idealism

CONCEPTS
- Idealism states that the idea, or consciousness, is the fundamental constituent of the universe. With the advent of information technology, the bit, the fundamental unit of information, has gained traction with other ontologies.
- Bits of information are compiled and stored in existence. Quantum superpositions and entanglement can be formalized as information transfer to yield new insights. Furthermore, there is strong evidence of physical-to-informational conversion in black holes.

IDEALISM, IN THE METAPHYSICAL sense of the word, has little to do with its ordinary usage (better to think of the word as Idea-ism). Metaphysical idealism holds that the universe is fundamentally made of consciousness, mind, spirit, or information.

Some of this is ground we have covered, a solipsistic hole that it is easy to fall into. Before the advent of systematic epistemology through measurable observation, experimentation, error correction, and explanation, direct experience was the whole of our scientific and philosophical purview. The expanse of existence was ascribed to a greater subjective consciousness, the mind of

God, and a hierarchy of ideas, worship, and even a papal organization was drawn up with eminence given to the scriptures. A construction of existence as the *world-as-experienced-by-God* was idealism's premiere showing on the metaphysical stage.

As the battle over the mind-matter problem gained momentum, some philosophers developed systems without the notion of physical substance. Included in these idealists is the Irish philosopher Bishop Berkeley who argued that everything we typically call physical objects are actually collections of ideas in the mind. All that we can ever know about objects is the idea that we have of them. Furthermore, while we are able to perceive of a few ideas at a time, there is a greater mind, the mind of God, that in His infinite perception is able to create and maintain both the ideas we think and the ideas we think of as physical objects.

This initial twist of idealism has been called immaterialism, the idea that the universe is solely made up of nothing more than thoughts that occur to you. Ultimately, immaterialism and theism gave way to an idealism based on the philosophy of Johann Gottlieb Fichte where "a transcendent, super-personal mind is the source of all aspects of the universe."[45]

Even with the depth of scientific understanding available today on the fundamental nature of the material world, idealism maintains two strongholds, one familiar and one new. First, that all of existence is actually the mind of God, a set of experiences that are meaningful but mysterious, especially as seen from our perspective—that of a small, insignificant subset of qualia in the deity's magical mind. The second argument that forms what could be termed *neo-idealism* stems from the meme of the computer age that understands the idea as information. More fundamental than an idea, yet still immaterial, information has taken the physics world by storm with many preeminent physicists looking to exchange *it for bit*, replacing matter with information.

> *The big bang at the beginning of time consisted of huge numbers of elementary particles, colliding at temperatures of billions of degrees. Each of these particles carried with it bits of information, and every time two particles bounced off each other, those bits were transformed and processed. The big bang was a bit bang. Starting from its very earliest moments, every piece of the universe was processing information. The universe computes.*[46]
>
> Seth Lloyd

Information exchange underlies, and in some cases, does a more thorough job of explaining the perplexing laws of quantum physics than does its physical counterpart. Idealism's claim to the fundamental nature of the universe is certainly on the mend when considering the informational components of our universe.

Qualia of God

While idealism is not necessarily a theological metaphysics, the long-standing argument that attributes *transcendental consciousness* to the personage of God is where we begin to develop our understanding of the properties needed to develop a logically consistent basis for the primacy of consciousness or at least a mental aspect to the makeup of existence.

In a recent meditation sit, I got to thinking of this idea of existence being in the mind of God and tried to be mindful of the *entities* that made up my mind. Most readily apparent are the sensations I have as the result of having experience largely embodied. Next most apparent is the often-present stream of thought, sometimes directed but often random and at all times capable of capturing all of my attention like the thorniest bramble. The ever-present first-person context in which these sensory or cerebral contents arise, consciousness, is the mirror in which all is reflected but is still "there" in some way which I imagine is much more robust with more experience in meditation. How would these general categories of entities be different within the mind of God? Which one are we?

In order to give thoughtful answers to these philosophical questions, it is important to dispense with the supernatural. If we consider the typical superpowers of God, where she is infinitely capable in manipulating both the contents and context of her mind, then the answer to the innerworkings of the mind of God is magic. No matter your question to probe the interesting concept of God's qualia, the deeper the dogma, the more fanciful the means to manipulate mind or matter. In all instances, a realm where supernatural metaphysics is considered is fiction.

Sans supernatural, the traditional concept of God as an entity with agency and power to act on his plan and our requests is surprisingly intact if we consider what it might be like living in the mind of some all-natural supreme consciousness. As we have already discussed, as a lowly evolved ape, I can have some pretty magnificent thoughts that consider beginnings, endings, meaning, processes, nothingness, multiverses, and incompleteness to name just a few; why should we consider the nature of our neighboring qualia within the mind of God to be anything less fanciful, appearing to us to be nothing short of miraculous? If this supreme consciousness had dreams, a persistent id, or was struggling to understand its own experience given whatever multidimensional realm it existed in, how likely would it be that as a pixel in this being's understanding of the color red or as a temporary bit player in its visualization of a 3-D world, our own experience would even register? Is the qualia that fails to draw your attention, the minor sensation or thought in your consciousness right now (maybe about a wrinkle on this page or white noise in the room), similar to our experiences' woeful impact on supreme consciousness?

Taking a step back, let us consider a couple of concepts in our own mind to illustrate the relative ease of considering a supreme consciousness, idealism's all-natural God, before we take her down a peg. First, let's take the brain-based explanation for experience. This materialistic argument says that slowly evolved, embodied feelings created new maps of information which were encoded on our brains (and to a lesser extent higher mammals and birds) and grew into our consciousness. All of this happens through the same feed-forward, feed-back recognition engine that allows us to move, read, or speak, namely as a function of the neurochemical processes in the brain. If the *ghost* of experience is run in the *machine* of the brain, then the gears and sprockets of the machine are the molecules of the neurons and neurotransmitters. These molecules follow certain processes based on their shape and composition.

The second concept, one that we will continue to return to for the tough questions it requires, is that of panpsychism. Panpsychism is the concept that all entities have some level of experience. Ants experience what it is like to be an ant. Atoms experience what it is like to be an atom. Granted, these levels of experience are likely not as profound as ours, given their memory restrictions, but it does philosophically close the case on the first question we want to ask any other articulate species, "can you experience?" (Since the second is, "Yeah us too... all the way down to suffering, so let's not do that to one another, capiche?) Panpsychism is a consistent philosophical stance considering the fact that no objective test exists to unearth the degree of subjective experience of an entity and a lack of mechanism where consciousness arises from some level of information processing... that is to say that without a robust understanding of what is conscious and how they became that way, it is more logically consistent to say that everything is conscious. The discussion for panpsychism always goes down and talks about earthworm experience, but far more interesting (and less grimy) is going up and considering if the universe itself is part of the panpsychic panoply.

Obviously if something as complex as the entire universe is on the subjective spectrum, we are not on the tail of the distribution that includes the "most" conscious of entities. Just as an earthworm's subjective experience cannot imagine the experience of the Dalai Lama, neither can His Holiness comprehend the experience of the universe. If the universe has extra-dimensions, supersymmetries, or parallel universes, it would also have *experiences of this existence.* Some of religions' most memorable speculations on God's experience, like the scripture that states "that one day is like one thousand years," are as underwhelming as the rest of the imaginings of pre-scientific nomads; if you think it is hard to discern if your wife sees the same color when looking at the millionth blouse in a department store, the difficulty of contemplating supreme conscious experience is infinitely worse. Until you can switch places with other entities *and* remain conscious of your own nature, possibly by plugging into a VR world like in "Being John Malkovich," you'll never know for sure, and moreover, it would not be the same as just being John Malkovich, it would be *you* being John Malkovich—a wholly different and problematic conscious experience.

However, we do not have to understand the experience of a supreme consciousness to dig deeper on the thought experiment; we just need to grant that inside a panpsychist's brain, each of the components have a consciousness analogous to how we have a consciousness and exist as a carbon-based lifeform inside the supremely conscious universe. What do you know of the atoms within the dendrites in your brain? Do their modest experiences impact yours? You might not think so, but we really don't know. If the atoms in question are in the neural plasticity formed during your most mindful or flow moments, and this is the majority of the time they are activated, is it their bottom-up excellence enabling your optimal experience? Top-down, do they feel special, chosen in some way? Does the analogy between us and our atoms break down because we are at the low-end of subjective experience and the idealistic universe is able to be mindful of its component's experience and its own? We have developed some clever experiments to find uncommunicative conscious components within our own brains and so we have some insight into what it might be like to have conscious subsystems and operating systems all in one mind.

Fig 13. - **Panpsychic Cog-In-the-Machine Meaning** - Assuming that everything has some experience sets humanity's consciousness up as a small part of a greater consciousness—the experience of the universe. We play a small but important role in this greater multidimensional consciousness, comparable to the role a single brain cell's panpsychic experience plays in your overall consciousness.

The first split-brain experiments were conducted by Roger Sperry and Michael Gazzaniga at Caltech beginning in the 1960s by asking a series of questions to severe epilepsy patients who had undergone a corpus callosotomy. By placing items in the left hand, which is controlled by the right hemisphere of the brain, subjects report not knowing the item, because the left hemisphere responsible for language processing is unaware of the item. Similarly, if presented with the word for the item in the far left of the visual field, the subject will not be able to verbally identify the word but will reach for the object with his left hand.

> *To the surprise of the first neuroscientists to conduct such experiments (and to the rest of us!), it seems that the same person can have two different answers to a question, along with completely different desires and opinions in general. And even more astonishing is the discovery that the feelings and opinions of each hemisphere seem to be privately experienced and unknown to the other. One "self" of a split-brain patient is as puzzled by the opinions and desires of the other as another person in the room would be. Whether or not both points of view in split-brain patients are conscious is difficult if not impossible to answer, but we have no reason to doubt that there is an experience associated with the thoughts and desires of each, and most neuroscientists believe that both hemispheres are in fact conscious.*[47]
>
> Annaka Harris

Clearly if our mind has separate consciousness in different hemispheres, there is a possibility that even smaller sub-systems in the hierarchy could also be conscious and just not able to communicate their level of experience to us in a way that we can understand. By analogy, subsystems all the way down to the experience of small organisms on planets might be part of the supreme consciousness of the universe and just be incapable of either transmitting or receiving communication with this higher quality of experience. We simply do not know enough about the nature of experience in either larger or smaller components of existence to say with assurance that consciousness only arises in systems like us.

The idealist's God, even stripped of superpowers, is an awesome entity capable of experiencing multiple dimensions and supersymmetries, an entity possibly aware of its sub-components' consciousness (e.g. us), and with agency to manipulate those lesser experiences but not the ability to communicate what it is doing with our mind either through experience or existence. Unless of course that is all that belief or faith in God is:

the weakly signaled experiential link between something abundantly more complexly conscious and those *tuned-in* to receive it. Like faintly hearing a lullaby in a language you don't understand above the roar of the seashore and still having the experience of wanting to sleep, there is no "knowing the *mind of God*," for the mind of God, just like any other mind, is only "knowable"—it is only *like something*—for that entity. But transmitting experience, even to subcomponent experiences, might be akin to reaching for an object even without being able to articulate why. The mind of the idealist God is split, fractured into a panpsychic myriad of sub-component consciousnesses, each only fully experienced by that entity, each largely unaware of its neighbors or its place as a single qualia in the ultimate mind of God.

Fig 14. - **Qualia of God Meaning** - Instead of focusing on the physical components of a panpsychic universe, the Qualia of God meaning imagines what it might be like to be an instance of experience, a *qualia*, in the consciousness of the universe. Think of your own stream of consciousness and imagine that any of those qualia could be an idealistic entity all its own. This meaning gets close to a modern conception of God who is coexistent with the "The Word" and can manipulate you, just as your imagination creates entire worlds when you are reading a suggestive book.

The panpsychic argument is a consistent metaphysics of experience given our current lack of understanding of the mechanisms and the proliferation of consciousness in *this universe*. Given the frame of only having this universe to deal with and knowing that both you and I are having rich subjective experiences (and likely so are our pets and some of the other more intelligent species around us), a logical speculation is that other organisms, that all entities—there is no logical reason to stop at just organisms—are having some level of experience. It would not change the laws of nature that govern existence, if the quantum particles that dominate the mechanisms of matter and energy had internal experiences, nor would it alter the consistent geometry-driven nature of the motion of stars, planets, and black holes if these too had a complex inner life. We should not expect that any of these entities could transmit their "hello, world!" proof-of-subjective-life to us.

However, if we consider there is more than one universe, it throws into question the necessity of adding consciousness to everything and allowing us to once again claim the more tangible ground that experience is resultant from some level of information processing *across universes.* As we have already seen, the nature of our experience of time is really an experience of the interaction between near-parallel universes; similarly, as we explore the distribution of conscious experiences we'll find they can also be best described as differing qualities of experience made more likely by their relative quantity of proliferation across the existence of the multiverse. In other words, we can begin to explain our consciousness as the distribution of experiences arising from information processing across parallel universes. We will dive more deeply into the nature of *multiversal consciousness* and investigate if these conceptions of existence help to solve the hard problem of consciousness in chapter six.

Even though the idea of multiversal consciousness does not require that everything be conscious, we are not done with metaphysical idealism yet. This section delved into the nature of a supremely conscious idealistic entity—the experiential speculation on the mind of God that has been the mainstay of idealism. In the next section we will probe the depths of a new form of idealism, to coin a term, neo-idealism, that largely comes from theoretical physics and computer science—an existence fundamentally made not of matter, but of information.

Bit by Bit

Information is a relatively recent discovery. At the same time as the semiconductor transistor was invented, in 1948 in the Bell Telephone Laboratories, Claude Shannon's work from the same institution, entitled "A Mathematical Theory of Communication" detailed the theory of what was fundamental to that first semiconductor transistor gate.

> *Information was found to be everywhere. Shannon's theory made a bridge between information and uncertainty; between information and entropy; and between information and chaos. It led to compact discs and fax machines, computers and cyberspace, Moore's law and all the world's Silicon Alleys. Information processing was born, along with information storage and information retrieval. People began to name a successor to the Iron Age and the Steam Age. "Man the food-gatherer reappears incongruously as information-gatherer," remarked Marshall McLuhan in 1967.*[48]
>
> James Gleick

Information, simply stated as the ability to answer a yes or no question, the ability to make a distinction, began to take the world by storm. As quickly as the technology was developed, the theoretical underpinnings raced to meet and often exceed it. Information's use in mechanisms enhanced its credibility. The theoretical underpinnings describing the bit and its use in computation revolutionized the world. Even with all of these accolades and technological spinoffs, information's fundamental role in the nature of existence is evidenced by its ability to recast the evolution of the wavefunction and quantum entanglement as information processing.

While all of the components of the standard model are shown as particles in their collapsed form, what we know from the Schrödinger equation is that they are actually a wave of existence much different from the defined, binary, *yea or nay*, existence observed in the macroscopic world. They are this probability wave without strict classical definition, that is, until they are observed, until a yes or no question is asked such as "are you here?" or "is your momentum five?" Our measurement tools do not disrupt the measurement in the classical way (a clumsy photon t-boning into our object beam), but the very act of measuring interactions in the quantum realm causes the collapse of the probability wave into a discrete particle, setting the stage for the famous Schrödinger's cat paradox.

Schrödinger's Smeared Cat

The most well-known paradox stemming from quantum physics is the paradox of Schrödinger's cat. This famous thought-experiment leaves us with a visceral sense of the mystery of the quantum world.

The setup is simple: a cat's fate, to live or die, is put into the hands of a quantum event—the decay (or not) of a single atom. The release of poison triggered by the decay of the atom (as measured on a Geiger counter) is the grisly instrument of the cat's fate. A closed but accessible peephole is provided for the sociopathic and scientific observer alike.

The decay of the atom is a quantum event, ruled by the wavefunction in the Schrödinger equation. A superposition of decayed, not-decayed, and decayed-not-decayed fully explains the state of the atom prior to observation, *before information is committed to what the state is*.

And this is fine for the atom, to be *smeared out* across these states; but what of the cat? Are we to believe that the impact of the measurement problem is the same on the macroscopic as it is on the single elementary particle? Why should our measurement—the request on the system to run an algorithm and answer definitively *dead or alive*—impact the outcome?

Erwin Schrödinger set up this thought experiment to point out the problem in the Copenhagen interpretation. The Copenhagen interpretation states that quantum systems remain in a superposition of all possible states, as described by the wavefunction, and is our only explanation for the quantum realm until a measurement is made. The Copenhagen Interpretation castigates us that the *uncollapsed* wavefunction has no physical significance.

The Schrödinger cat thought-experiment is now utilized to understand the consequences of different physical theories: the many-worlds interpretation of quantum mechanics results in a "new" universe for each quantum decision (one for alive cats, the other, sadly, for dead ones); decoherence says that since all of the participants (the cat, the Geiger Counter, the decaying atom, and the observing human) are made of quantum material they evolve in a complex wavefunction together; and finally, another approach gaining momentum is that the universe's algorithm calculated a solution and the information, the answer in the form of a particle, was displayed. It is this algorithm's experimentally well-documented tendencies that we know as quantum field theory.

> *And in this age of computers, and information, and flashing pixels there is nothing counterintuitive about the ontological idea that nature is built—not out of ponderous classically conceived matter but—out of events, and out of informational waves and signals that create tendencies for these events to occur.*[49]
>
> Henry Stapp

```
The scene is of Heisenberg and Schrödinger
```

driving down the highway at a prodigious rate of speed. As they pass mile marker thirty-four, a state trooper hits them with a radar, chases them down, and pulls them over.

The trooper walks up to the driver's side window and finds Heisenberg behind the wheel.

Trooper: "Sir, do you know how fast you were going when you passed that mile marker?"

Heisenberg: "Well, I cannot because I now know where I was."

The trooper detains them and does a search of the car. In the trunk, he finds a dead cat. He goes back to his car where the two arrestees are sitting in the back seat.

Trooper: "Did you know there is a dead cat in the trunk of your car?"

Schrödinger: "Well, now I do."

IDEALISM

On the Seventh Day He... Collapsed Wavefunctions

Utilizing either the explanation of decoherence—that the quantum interactions between measurement device and the quantum event under investigation eliminates the paradox of superpositions in the macro-world (i.e. the alive-dead cat)—or the continuous evolution of wavefunctions with collapsing universes as in the many worlds interpretation of quantum mechanics, the heavy lifting expected of conscious observers in lab coats is greatly reduced. In other words, wavefunctions are not "waiting around" for someone to observe them; rather, interactions with other nearby quantum systems are enough to enable these collapses.

But how should we explain these interactions? As we have seen, extreme idealists like Bishop Berkley have claimed that the existence of everything from the unobserved moon to distant wavefunction collapse required God's mind's eye to observe it—this obviously is not what we are looking for. Instead, neo-idealists believe the wavefunction evolves in a similar way to how the information processing and data retrieval in the semiconductor memory in your computer works. Essentially, when a current is passed on one end of a row of transistors, the information of "shorted" (on) or "open" (off) is requested. The row of transistors is pre-programmed (or not) with individual electrons that either cause the short, or if they are not there, enable the open.

You are responsible for the programming. When you save a song like "Things Have Changed" by Bob Dylan, 7.7 MB or 61,600,000 individual transistors or bits are programmed. It is information forever stored on transistors that will result in either "shorted" or "open" channels under strings of these transistors. In the most modern (2015) floating-gate transistors, less than twenty electrons reside on any individual transistor. Your programming uses a quantum phenomenon known as tunneling, the non-classical movement of electrons from the channel onto the gate, *through* the insulating layer of silicon dioxide. The particular form of quantum tunneling used in floating gate programming, Fowler-Nordheim tunneling, is not entirely random and is *motivated* by a high electric field.

So, when you begin the song and hear the lyric, "a worried man with a worried mind, no one in front of me and nothing behind," the engineering of numerous quantum information processes is involved. First, we know that we cannot know that an electron is *on the gate*, it is the classic *particle-in-a-box* problem and its wavefunction is *smeared* from bit-line to channel. Error correction is required to overcome this and other manufacturing quality issues (that are far more prevalent) and if transistors are not performing correctly, they have backup. Furthermore, the channels' "short" or "open" states are not always perfectly defined, so the signal must be amplified for the processor to give you studio quality enjoyment of the "Poet Laureate of the Blues."

In the case of your flash memory solid state device (SSD), numerous measurements are designed by engineers that engage with the quantum wavefunction collapse. These designs were integrated into the circuitry in order to read, write, erase, amplify, and correct for quantum events. Furthermore, many of the items you programmed (that Zamfir song you downloaded on a lark) will remain unmeasured and yet accessible to a randomization of your entire playlist, a measurement made utterly by chance.

The measurement of quantum processes on your local SSD memory are only *observable* by you or your local area network (LAN), but the streaming service of the same song is shared by as many (or as few) users as are interested in that particular arrangement of binary information, the encoded songs.

> *The observation... enforces the description in space and time but breaks the determined continuity by changing our knowledge. The transition from the "possible" to the "actual" takes place during the act of observation. If we want to describe*

> *what happens... we have to realize that the word "happens" can apply only to the observation, not to the state of affairs between two observations.*[50]
>
> Werner Heisenberg

The universe's processor and SSD circuitry appears to be arranged in a similar way. There are "wires" leading out from the quantum world to the macroscopic that output something about the program and the states of information. Physical *stuff* is formed only when information is accessed by the networked end-user in the same way that the song plays when information is requested from deep within the transistor stack's information of "opens" and "shorts." We have been so used to relying on the explanation of tiny stuff that changing our frame to packets of data at the fabric of the natural world is difficult, but this frame is useful for some of the most upsetting paradoxes of the quantum world, especially the idea of entanglement, what Einstein called "spooky action at a distance."

Spooky No More

The quantum property of entangled states helps us understand how information is integral to our universe. Entanglement involves communication between quantum particles that keeps their properties—like spin—in sync. This occurs even at distances where this instantaneous communication would have to take place at speeds greater than the speed of light. Quantum physics requires entanglement in order to maintain the Heisenberg uncertainty principle, which states that we cannot simultaneously measure both the position and the momentum of a particle. The famous Einstein-Podolski-Rosen thought experiment shows the necessity of entanglement in quantum theory by considering the behavior of two wavefunctions generated by the same source but propagating in opposite directions. Let us say we decide to measure wavefunction L (moving left) for momentum, in which case we are treating it as a wave and measuring its wavelength. When we look at wavefunction R, we correctly assume it has the same magnitude for its wavelength, just in the opposite direction. This means wavefunction R is also collapsing into a wave. If instead, for wavefunction L, we chose to measure its precise position, a particle resolves and its entangled particle would come from the collapse of wavefunction R at the same magnitude of position, but to the right of the generator. There is no way to trick the system, measuring position results in only information on position from the collapse of both entangled wavefunctions into particles. Quantum entanglement is the process of nature that enables the uncertainty principle since measurement provides *only* information on momentum *or* position for an entangled pair of particles.

Scientists have constructed experiments around the Einstein-Podolski-Rosen (EPR) thought experiment that show correlation between entangled particles. Bell's theorem took the theory of entanglement from the philosophy classroom to the experimental laboratory. In 1982, a team of physicists in Paris, tested Bell's theorem and the EPR equations and found that two photons emitted from a calcium atom were correlated to an entangled degree.

> *When particles or quantum systems are entangled, their properties remain correlated across vast distances and vast times. Light-years apart, they share something that is physical, yet not only physical. Spooky paradoxes arise, unresolvable until one understands how entanglement encodes information, measured in bits or their drolly named quantum counterpart, qubits. When photons and electrons and other particles interact, what are they really doing? Exchanging bits, transmitting quantum states, processing information. The laws of physics are the algorithms. Every burning star, every silent nebula, every particle leaving its ghostly trace in a cloud chamber*

is an information processor. The universe computes its own destiny.[51]

             James Gleick

A hypothesis that there is a vast network of information accessible to each quantum event is one way to make sense of worlds-apart entanglement and superpositions of being and nothingness. Neo-idealism is based in information theory's solution to entanglement: interactions we see as material outcomes are fundamentally the universe's way of answering *on* or *off* for the qubits it has at its disposal. The actions of entangled particles are no spookier than accidentally sending two left-handed gloves to your mom on Mother's Day and her *instantaneous, accurate information* that you have two right-handed gloves still to send her.

> *Physics experiments can bind the fate of two particles together so that they behave like a pair of magic coins. If you flip them, each will land on heads or tails— but always on the same side as its partner. They act in a coordinated way even though no force passes through the space between them. Those particles might zip off to opposite sides of the universe, and still they act in unison. The particles violate locality— they transcend space.*[52]
>
>   George Musser

Bits from Black Holes

Black holes are studied by cosmologists to understand the interplay of quantum mechanics and relativity. A black hole contains many of the extreme conditions of the early universe—a singularity contained inside a claustrophobic spacetime well. Our hunt for the physical evidence of informational idealism's circuitry continues with these inexact but useful geometric models of the early universe.

Black holes were found first in the math of relativity. Einstein calculated the formation of an infinite curvature of spacetime from a collapsed star but did not live to see black holes verified by their gravitational effects on nearby entities and at the center of galaxies. While black holes took a backseat in theoretical physics to studies in quantum physics, the world waited for Stephen Hawking and Roger Penrose's important realization that black holes contained a singularity similar to the one responsible for the big bang. However, an even bigger paradigm shift occurred not from thinking about the singularity at the "bottom" of a black hole, but the entropy at the surface, for these large randomizers of both matter and information help to link the world of *it* (matter) and *bit* (info) in a very elegant way.

The second law of thermodynamics states that entropy (disorder) in a system is always increasing across all localities. Imagine a loaf of baking bread. The odors made up of many gaseous molecules do not stay inside the oven but escape and diffuse throughout the entire house. It is especially unlikely that the molecules would arrange themselves in such a way as to avoid your nasal cavity or stay packed within the bread. The molecules' diffuse nature upholds the second law. This physical law— entropy has to be increasing—must apply even near a black hole.

In an apparent contrast to thermodynamics, masses falling into a black hole are compressed and go from being complex starships to tightly wound singularities, the antithesis of disorder. However, as Stephen Hawking and others began to explain different interactions at a black hole, a more complex picture formed that allowed entropy to not only be conserved but also translated from matter into pure information.

Hawking theorized that instead of matter thrown into a black hole being translated from potential to useless energy like is evidenced in most thermodynamic reactions, the matter's entropy was conserved in a more fundamental way—translated from the data that you have (those overall macroscopic features) to

IDEALISM

the data that you don't (the system's particular microscopic arrangement).[53] Much like matter can be broken down into its constituent particles, information can be divided down to its smallest unit, the bit, the ability to answer a yes or no (a 1 or a 0) question. A bit is a single answer, hence a single unit of information.

In order to illustrate, let us take as an example the information of a coin. We'll first define the bit as the ability to answer a yes or no question, or in our system of coins, a heads or tails question. Our system of 1,000 dimes, arranged in a neat circle, one against another, as tight as we can arrange circles within a larger circle, contains 2^{1000} different possible configurations but only 1,000 bits of information, 1000 answers to the question "heads or tails?" Even if we know none of the microscopic details of which side of the coin is facing upwards, we still know there are 1,000 units of entropy, one for each dime.

Now imagine we added an unknown number of additional dimes to our circle, still unsure of the answers on each new unit of information. The gap in our information is proportional to the increased size of the surface area covered by our dimes. To restate, the configuration of the data *and* number of bits may be unknown, but it is related to the surface area of the dime database. A blackhole stores information in the same way, just at a much greater density.

Once a black hole is formed, matter captured in its gravity is translated into information and the surface area of black hole increases.

> *Hawking was not easily convinced, however, and so over the following two decades physicists developed a new theory that could account for the entropy discrepancy. This is the holographic principle, and it holds that when an object falls into a black hole, the stuff inside may be lost, but the object's information is somehow imprinted onto a surface around the black hole. With the right tools, you could theoretically reconstruct*

> *this magazine from a black hole just as you could from the pulp at the recycling plant. The black hole's event horizon—the point of no return—serves double duty as a ledger. Information is not lost.*[54]
>
> Michael Moyer

By dividing up the surface area of a black hole by the entropy it contains, Hawking and others found that each unit of information was set in a square of Planck's length on either side. As a bound for the information storage, this is a very small area; but since Planck's length is a fundamental constant of the universe, the smallest unit of distance thought to make sense, there is further corroboration for idealists to state that information is the fundamental constituent of existence.

> *But the physical environment itself is emergent; it arises from the fundamental ingredient, information, and evolves according to the fundamental rules, the laws of physics.*[55]
>
> Brian Greene

This result affirmed the second law of thermodynamics in a black hole and bound information onto a fundamental unit of surface area. Scientists went on to the similarities between the singularities of black holes and the big bang to project the principle of surfaces of information onto the fabric of the entire universe. In other words, the expanding surface of a black hole that contains information on all the stuff thrown into it is a smaller version of our universe where the surface of the universe contains the information of all that exists within the bulk. For the idealists, the surface of the fabric of existence is not empty anywhere, instead everywhere makes up the biggest data center ever comprised. The surface of our universe is a 10^{122}-bit hard drive[56].

Our understanding of the quantum world is best explained by considering what we observe and calculate not as particles or waves, but as information. As Gleick points out, "why does nature appear quantized? Because information is quantized. The bit is the ultimate unsplitable particle."[57] In the next chapter, we will look deeper into the physical theory that most compellingly explains how the universe uses the information on the surface of its fabric to calculate and project our reality—a material holograph.

Scientific life after death?

Idealistic metaphysics suggests four separate routes for meaning: two we have discussed where we interact as a subcomponent or single experience in the panpsychic mind of a supremely conscious universe[58]; the third, slightly different, where everything is conscious but there is no experiential interaction between subcomponents; and forth, a universe fundamentally made of information. Let's cover the last two in turn.

As mentioned in the previous chapter, our subjective inquisitiveness spins a very rare story into a universe without its own meaning. This inside-out universal meaning is the widescreen setting of physicist Sean Carroll's *Big Picture* poetic naturalism[59]. Panpsychism goes a step further and universalizes some gradient of consciousness into all entities in existence, eliminating our unique poetic role but granting the meaning that comes from some appreciation of experience to all of existence. Even if even quarks have some appreciation of experience, it would not answer the question of if meaning is inherent to the universe. Meaning in this case would still be inherent to consciousness, but thanks to the universal attribution of consciousness, the panpsychic universe would approach universal meaning from the prevalence of "personal" meaning.   If everything is to some extent conscious, then there are just greater and lesser degrees to which meaningfulness can be experienced. Meaning would be *personal* but prevalent.

Prevalent-personal meaning is an important concept that we will return to in subsequent chapters when our frame broadens to prevalent-personal meaning interference across many parallel universes, but for now, we are just considering it within "this" universe alone... for all entities in a panpsychic system. Prevalent-personal meaning is universal meaning "within the limit" to borrow a term from mathematics. As the panpsychic universe includes consciousness for all entities large and small, the addition of each thing's meaning to the pile approaches truly universal meaning, the universe is completely filled with personal meaning. While anything "approaching the limit" of a concept is not truly the concept, the difference can be insignificant. However, in the case of panpsychic prevalent-personal meaning, I do not think approaching the limit makes for universal meaning, especially when considering multiversal prevalent-personal meaning which we will discuss in subsequent chapters.

First, most people balk at the absolute that attributes consciousness to all entities in existence. Our intuition says that if electrons or silicon or slugs have some gradient of consciousness, they would present themselves in a different way. We certainly can't prove or disprove this intuition with our current scientific understanding of consciousness and its constituents. Our current inability to test the degree of consciousness in our computer systems belabors this point. However limited our knowledge of consciousness's constituents, it is very likely a function of information processing, as is evidenced in us by the different look of FMRI brain scans between the complete unconsciousness brought on by anesthetics and the largely unconscious nightly sleep. A slug or a grain of sand would require a different route to its panpsychic consciousness than we attribute to our complex modular brain. So, while we can't disprove the panpsychic prevalent-personal meaning, it is largely argued for consistency's sake for *this universe*. We will see later, as we add parallel universe interference to the mix of consciousness, panpsychism is unnecessarily bloated.

IDEALISM

The second reason I do not think panpsychic prevalent-personal meaning approaches universal meaning is that there is no connection between the personal meanings of the various constituent entities, no means for them to create a synergy of meaning, or even understand if it is like something to be another type of entity. It is not necessary to broach personal meaning barriers for there to be a *type* of universal meaning, but this would be akin to changing the definition for universal meaning mid-stream. It is better to call this type of meaning what it is, prevalent-personal meaning, and add it to our litany of possible forms of meaning that may exist in the universe—one that without some form of connection *between* nodes cannot rise to the level of universal meaning as we have thus far defined. The connection does exist, and we will explore in great detail the interference of prevalent-personal meaning *across* parallel universes in subsequent chapters.

The fourth form of universal meaning from the idealistic universe, is the universal meaning of remembrance.

An important form of personal meaning happens when we take solace in the remembrance of our loved ones that have died, of the good times that we have had with those that have moved far away, of our youth as we grow older. We add meaning to these events, often wanting to live more in the meaning-imbibed past than in the painful present. For all of human history, we have built on the comfort of our memory of times past and our remembrance of loved ones who have died by constructing realms that offer the type of metaphysical meaning most desired in human history—a life beyond death—a place where our remembrances are forever reenacted.

As an atheist living in Mormondom, I am confronted by the unique human need for solace for those that have been left behind. In both of the neighborhoods where I have owned homes in Salt Lake City, I have befriended elderly women that are not religious (in a theocracy the non-believers often stick together!), so I am close enough to them to talk frankly and have been glad to offer my consolations when each of their husbands passed away while I was their neighbor.

One claim that I am unable to make is that their husbands are "in a better place." While I am openly glad that they have passed out of the suffering of protracted illness, this is adding a silver lining to the moment that is upsetting to my lovely neighbors. My consolations fail to address either the end of personal meaning for the deceased or my neighbors' increased suffering from not having that person around. I listen and am understanding, but when asked if I *really* think it all ends with death, I can see the impact of my answer in the affirmative—they are crestfallen.

An afterlife is at once the most desirable and most undesirable component of religion. The purported paradise on the supernatural plane of existence that we go to when we die offers the solace of a reconnection and continued generation of meaning in the afterlife, but offers a mixed bag of behaviors among those believers living for the ever-after in the here-and-now. Some perform good works to get into the good place, bettering all of our time here; but others claim a scriptural fast pass for themselves and their tribe, relegating all others to the bad place, and misbehave accordingly.

A universe of information, not even necessarily experiential or conscious, but a huge hard drive that reads, erases, and writes existence in some or all of its granularity would have some record of all of our interactions, even after we died. Our life's existence is one bit of *being*, and our bit of info is our own chalkboard. One of the black ones, one for us alone. Our bit is a single scratching, as though leaning and thinking against a cool piece of graphite, ready to write. Painting an intellectual masterpiece, onto your blackboard opus, is the most soulful and consequential epic in which to focus our bit-like (but universally accessible) examined life. Our life written on the fabric of reality could be recorded and meaningful long after we have ceased to exist.

IDEALISM

It is hard to know how much credence to put into this sort of universal meaning. Putting this idea to the test with my mourning neighbors is an experience replete with blank stares and the unmooring of conversation. The idea of using the neo-idealist universe as a means to store the lives and meaning of our loved ones, a sort of *heavenly database*, is a novel and not at all subscribed idea. The major gap is reading the information. Whether the encryption of the universe's information storage of those we love that have passed away—or any other information we might want to directly access—is made impossible by the laws of nature or just a distant technological and theoretical leap for us to access is a question for future researchers.

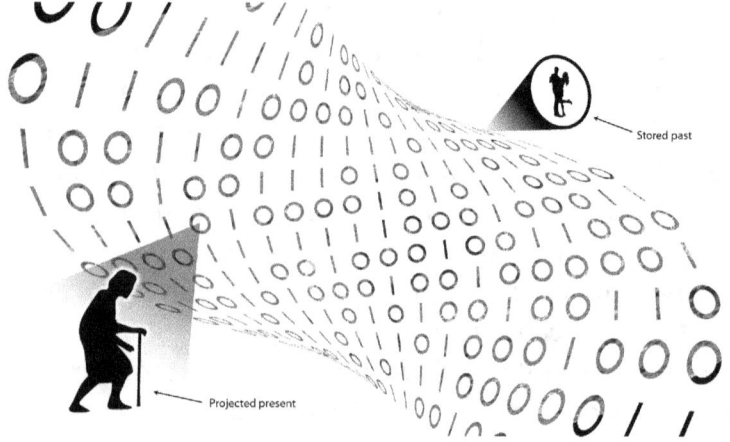

Fig 15. **Heavenly Database Meaning** - The holographic universe is projected from a quantum mechanical database on its surface. This database is capable of also storing information of the past in its vast data banks. The right process or technology might be able to defragment this data and project our past into a near-parallel universe.

Idealism has run the gamut from the mind of God to the information manifest in the universe. Now more than ever, idealism and materialism stand on nearly equal physical footing, a stalemate between mind and body, between information and matter that begs for a negotiation. The unifying theory that answers both the philosophical and physical questions on the nature of existence and our place in it springs forth from the intellectual firmament sewn by this epic struggle.

The hammer that will break down the Cartesian wall that separates these metaphysical camps has its seeds in the latest theoretical physics. This new process relational ontology compiles the idealists' information into the stuff of materialist quantum field theory and relativity. Without adding further realms of supernatural or even unverifiable metaphysics onto the fire, a computational multiverse clarifies the world, our experience of it, and its meaningfulness.

CHAPTER GLOSSARY
metaphysical idealism - the mind of God, the idea, or consciousness is the fundamental constituent of the universe.

neo-idealism - the bit of information is the fundamental constituent of the universe. The universe is an information processing and storage medium.

bit - the smallest unit of information, the measurement of a distinction often between on or off, 1 or 0.

qubit - a quantum bit which has three possible states, the classical bit's 1 and 0, and the quantum superposition of these states defined by the wavefunction before collapse, 01.

entanglement - apparent faster-than-light communication between particles that Einstein called "spooky action at a distance."

If entangled properties (like particle spin) are considered informational outputs of a computational universe, the apparent speed limit breaking paradox is eliminated. Put another way, interactions we see as material outcomes are fundamentally the universe's way of answering on or off for the qubits it has at its disposal.

KEY TAKEAWAYS
- Idealistic ontology began as a largely dualistic one that purported that all was the *mind of God,* that the universe was part of a greater supra-consciousness. Consciousness or *the idea* was seen as fundamental and all of creation was emergent from, and beholden to, God's perception. A panpsychic universe, where every entity is conscious, places us as either a physical or qualia sub-component in an all-natural supremely conscious universe. Our linkage to the universe would be both physical and experiential.
- In the mid-twentieth century, a new fundamental component—information—was suggested first in mathematics and later into idealism. The utility of framing newfound quantum physics paradoxes as information exchanges helped set into motion the invention of semiconductors, the internet, and computation.
- The surface of the universe is theorized to contain all of the information necessary to *project* our holographic universe. The fabric of the universe is a database of bits, a web of informational nodes in service of the computation of our universe.

CHAPTER 5:

Process Ontology

CONCEPTS
- Process ontology claims that processes (wavefunctions, optimization, consciousness, etc.) are the fundamental constituents of the universe. The laws of physics are part of this greater computational universe.
- Quantum gravity theories like holographic universe theory describe the combination of all the fundamental forces at the big bang and are making important mathematical discoveries by translating from a realm of quantum computation to a materialistic realm.
- Virtual reality can replicate existence with high fidelity to the extent that it utilizes the known laws of physics in its computation. Taken to the extreme of computational efficiency and memory capacity, the known universe is utilizing quantum computational processes to create reality.

PROCESS-RELATIONAL PHILOSOPHY, AND THE computational universe it suggests, offers a synergy between the materialistic and the idealistic metaphysics covered in the last two chapters, a compelling explanation for why existence is comprehensible (that is, why it appears to run by a small subset of rules that we can ascertain through conjecture, experimentation, and error correction), and a means for us to talk reasonably about a universal optimum, a subroutine of meaning being actively run in the reality of the cosmos. In this chapter, we will use computational terms like algorithm and compilation interchangeably with more common terms like process or system. The formerly abstract laws of nature or physics will be ejected from the connotation of existing as a static representation in a book into their true place in the universe—as the code running the whole show. Process metaphysics speculates that we live in the Matrix where code is "behind" everything and the cast of characters are material and informational (not a malevolent father-architect nor the Puckish mother-oracle).

> *Process philosophy is an effort to think clearly and deeply about the obvious truth that our world and our lives are dynamic, interrelated processes and to challenge the apparently obvious, but fundamentally mistaken, idea that the world (including ourselves) is made of things that exist independently of such relationships and that seem to endure unchanged through all the processes of change.*[60]
>
> C. Robert Mesle

The Way

Process philosophy is ancient. First penned in the sixth century B.C. by Lao Tzu, the Tao Te Ching both illuminates and obscures Tao, roughly translated as "The Way."

> *The Tao is like a vase that is empty yet used. It is the emptiness that gives birth to the vase. This emptiness, deep and unfathomable, is the source of the ten-thousand things.*[61]
>
> Tao Te Ching

Taoism (pronounced DOW-ism) is a theory of existence that maintains that what is ultimately real is not *stuff* but rather a process. The Tao is the non-being responsible for being. Your cereal bowl is an excellent example of the utility of non-being, the space the bowl provides is inexhaustible. Taoism shows that the transcendent attainment of understanding of the Way is only possible by balancing yin and yang, finding the middle path of deliberate practice, while ensuring our actions and service in the real world are maintained. Process through-and-through, the Way is neither matter-energy nor idea-information, the Way is neither being nor nothingness. The process that translates it to bit and back, that enables being and distinguishes nothingness, is Tao.

> *Unless the mind, body, and spirit are equally, developed and fully integrated, no [wisdom] can be sustained.*
>
> *When the mind and spirit are forced into unnatural austerities or adherence to external dogmas, the body grows sick and weak and becomes a traitor to the whole being.*
>
> *When the body is emphasized to the exclusion of the mind and spirit, they become trapped like snakes: frantic, explosive, poisonous to one's person. All such imbalances inevitably lead to exhaustion and the expiration of the life force.*
>
> Hua Hu Ching

The parallel between the code in *The Matrix* and the Tao, the process that results in the creation of everything in this simulated universe, and ultimately, even in "the real world" can be seen throughout the series. Neo's transcendence as *the One* is complete when he sees everything in terms of its encoded process.

PROCESS ONTOLOGY

The Matrix is everywhere. It's all around us, here even in this room.

Morpheus, "The Matrix"

The Tao is hidden but always present.

Lao Tzu

Western philosophy saw its first foray into process ontology with the Greek stoics. They contended that "the material world [is] pervaded by a dynamic force, which acted not mechanically but purposefully, in order to maintain a universal rational pattern throughout nature."[62] This force does not come from outside nature but is the motive force behind (and a fundamental part of) nature. For the stoic, the physical laws are rational and (in order to avoid fatalism in a highly-determined world) filled with goodness. Unlike in Platonic idealism, there is no means to interact with this rational process. Stoic suffering arises from living in discordance with the predetermined events humming along to the purpose of the universe—far better to accept and believe in the benevolent, all-natural, transcendental, empowering process behind the movement and manipulations of the material world.

With few exceptions, our view of the world since Descartes has been one of either stuff or ideas. The most recent example of process relational philosophy's attempt to negotiate a settlement between mind and body dualism is the philosophy of Sir Alfred North Whitehead. Contemporary advances in relativity and quantum theory propelled the work *Process and Reality,* in which Whitehead contributes the first new serious metaphysical ontology since the Cartesian split.

Whitehead challenges Kant's claim that existence does not have space, time, and causality and that we can know this because we experience ourselves as part of the larger causal world. Whitehead argues that we do experience causality both at the sensory experience level and at a deeper level, essentially saying that our *process of becoming* is highly related to those of the near-space and recent-time processes of a computational universe[63].

The subtle reframing of materialism and idealism that places process ontology as a likely candidate to offer a consilience between these two great pillars of metaphysics is also occurring in physics where calculations are showing the algorithms of nature to be the most fundamental constituent of existence.

Based on his studies of the properties of black holes, Stephen Hawking postulated the first signs of a viable truce between matter and information. It is here that the universe most decisively exchanges matter for information. The act of throwing a book into a black hole and accounting for the entropy increase required to satisfy the second law of thermodynamics leads us eventually to our third metaphysical ontology and a new candidate vying for our universe's most fundamental constituent—the system of related processes. For the remainder of this chapter we will discuss the two candidates for the "operating system" of the processes at the foundation of existence: holographic and quantum computational processes.

The Universe in a Funhouse Mirror

We have come to a negotiated end of the war between stuff and info. Both have a right to stake their claims to large swaths of consciousness and the cosmos, and certainly further details can be understood about each in their domains of current research and technological development, but neither appears to be fundamental. Something that explains both matter and information, experience and existence, may offer a more complete theory. A computational universe turns a funhouse mirror on matter and information and compiles the laws of physics and the universal constants into our physical world. One of the leading computational universe models compiles rules and information on the surface of the neo-idealistic universe into a physical manifestation (the materialistic universe) utilizing a process similar to the one which projects a hologram from a special film. This holographic universe is the most detailed and dynamic hologram ever imagined, and its fundamental component, the process of projection, changes the way we think about the universe and our place in it.

My first contact with holograms, like with so many other things, came from baseball. More specifically, my baseball-card collection where faux-holographic cards were marketed to unknowing kids and their placating parents. They were very neat, the players throwing and then contracting, swinging and then trying to check their swing. Angle the card faster and faster and the starter wears out his arm having to both throw and retract his pitches. Methodically experiment to optimize the angle of the card to resolve both images to have equal clarity or equal effervescence. Later, my university had a true holographic display, and I was hooked after I committed civil disobedience by putting my hand through what otherwise looked like an apple.

Holography is a technique that enables a field to be recorded and later reconstructed.[64] A hologram is made on a recording medium similar to those found in photography (a photographic film of silver halide photographic emulsion with many more light sensitive receptors than in photography) by directing one part of a split laser beam toward the object—the image you desire to make a hologram of—and the other half of the laser beam at the recording medium. When displayed, the recording medium does not look like anything special; you cannot tell what image is recorded (similar to how it is impossible to perceive what is recorded on a DVD just by looking at it), but when another laser is directed at it, it displays the entire contents of the object originally recorded. While most of the holograms used in security, memory storage, or art are made from light fields, like those from lasers, it is theoretically possible to use other types of fields to produce holograms.

Holograms have some very interesting properties. They represent a capture of the entire object from all vantage points, presenting themselves as three-dimensional representations of the original object. Furthermore, it is not that a one-to-one correlation of recorded image to projection is needed—instead any of the hologram's numerous photosensitive regions can record, and is capable of projecting, the entire image.

A hologram is formed through the physical principles of diffraction and interference. Diffraction is the effect of waves scattering after striking an object, while interference is the modification of a wave when another wave comes into contact with it. Interference can either be constructive (when two waves are in-phase and make a bigger wave) or destructive (when two waves are out-of-phase and shrink to a smaller wave, or no wave at all). When the illumination beam strikes the object and the reference beam strikes the recording medium they diffract the laser beam. The interference between the illumination and reference beams determines the properties of the recording when the interference wave arrives back at the recording media. The interferences create a particular change in the recording media where the phase was changed by the diffracted object beam, and none where there was no object to interfere with the reference beam's escape. When a similar light is diffracted off the recording media at a later time, it reproduces the recorded interference pattern of the object on all sides, even though the object has long since disappeared.

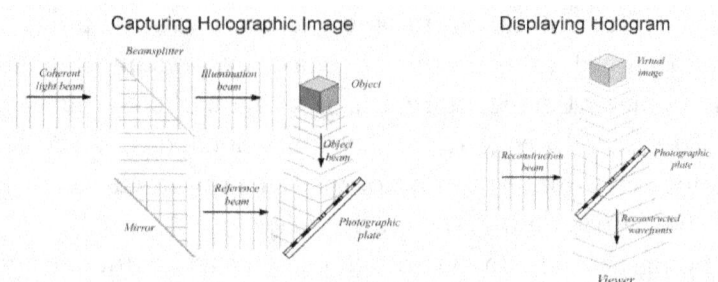

Fig. 16- The process for capturing and displaying holographic images

In essence, the holographic universe theory postulates that our four-dimensional universe (and everything and everyone in it) is *projected* from the information on the surface. Our universe is a hologram created by the projection of the information written on the holographic film on its surface.

> "In Plato's parable of the cave, our senses are privy only to a flattened, diminished version of the true, more richly textured, reality. Maldacena's [Holographic Universe] flattened world is very different. Far from being diminished, it tells the full story. It's a profoundly different story from the one we're used to. But his flattened world may well be the primary narrator."[65]
>
> Brian Greene

Holography offers one sort of computational universe, a universe dependent on the process that projects the information that is the fundamental fabric of the universe into the material fundamental in the bulk. These processes are not unknown to us; the computations of this holographic projection are the laws of physics, the equations from quantum physics and general relativity. Holographic quantum gravity is one of the most promising unifying theories. It is also the first to combine matter, information, and computation using the laws of physics.

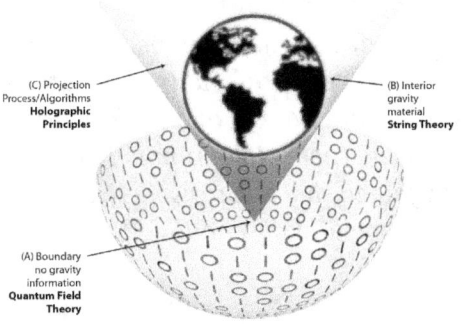

Fig. 17 - The Holographic Universe describes a process and informational universe where the information of quantum field theory on the surface of the universe (A) are compiled and projected (C) such that the interior material universe is governed by string theory (B).

Brian Green states, "the holographic universe may thus be more than a potential outgrowth of fundamental laws; it may be part of the very definition of the fundamental laws." In other words, it is not the stuff or the information that is fundamental, but the way in which they are translated that is important—the process. Furthermore, this process is intelligent and capable of modification to suit a set of initial conditions and, as we will see, an optimum purpose.

> *Since the information required to describe physical phenomena within any given region of space can be fully encoded by data on a surface that surrounds the region, then there's reason to think that the surface is where the fundamental physical processes actually happen. Our familiar reality, these bold thinkers suggested, would then be likened to a holographic projection of those distant physical processes.*

> *If this line of reasoning is correct, then there are physical processes taking place on some distant surface that, much like a puppeteer pulls strings, are fully linked to the processes taking place in my fingers, arms, and brain as I type these words at my desk. Our experiences here, and that distant reality there, would form the most interlocked of parallel worlds. Phenomena in the two—I'll call them Holographic Parallel Universes—would be so fully joined that their respective evolutions would be as connected as me and my shadow.*[66]
>
> Brian Greene

Reframing the universe on the process between material and information has already allowed physicists to do groundbreaking work. Holographic universes modeled with different mathematical spatial constructions than our own, known as anti-de Sitter space, have been utilized to translate quantum field theory *on the fabric* into string theory *in the bulk*. Doing these calculations to transform one brand of physical laws on the surface into another in the bulk has been extraordinarily helpful in progressing both theories. A physicist simply chooses the easier realm to arrive at an answer, completes the calculation, and transposes (holographically) that answer back to the original realm. While we have not yet been able to perform holographic calculations in the more complex spacetime we live in, these calculations have been of great utility and reveal a path forward to further develop the theory.

> *The important point from a quantum gravity angle is that the boundary theory [the one used on the surface of spacetime] is a well-understood quantum theory of particle physics, very similar to the one we use to describe sub-atomic processes in nature. Referring only to small scales, it doesn't bother with gravity. Yet it is able to describe the esoteric quantum gravity theory which governs the interior. It's the first ever complete description of a quantum spacetime.*[67]
>
> Brian Greene

Proof of such a profound model of the universe, one capable of unifying the four fundamental forces of nature, explaining the compilation of information into matter, and linking quantum physics and relativity, is underway. Leonard Susskind, a preeminent theoretical physicist at Stanford University, has reviewed papers by Yoshifumi Hyakutake of Ibaraki University in Japan and stated that, "they have numerically confirmed, perhaps for the first time, something we were fairly sure had to be true, but was still a conjecture—namely that the thermodynamics of certain black holes can be reproduced from a lower-dimensional universe." Maldacena notes, "The numerical proof that these two seemingly disparate worlds are actually identical gives hope that the gravitational properties of our universe can one day be explained by a simpler cosmos purely in terms of quantum theory."[68]

Experimental physicists like Fermilab Center for Particle Astrophysics Director Craig Hogan are also looking into measuring the projection directly, through something called *holographic noise*. According to reporting by *Symmetry Magazine*, the 2009 GEO600 experiment searching for gravitational waves emanating from black holes was plagued by unaccountable noise. This noise could, in theory, be a telltale sign of the universe's discrete nature, a web and not a continuous sheet, whose nodes are holographic quantum bits. Using a more precise laser interferometer known as a holometer, experimenters seek to measure spacetime with far more precision than any experiment before—and potentially observe effects from those fundamental bits at the surface of spacetime.[69]

> *Yes, you could say [we are] an illusion, or an emergent phenomenon,"* says Maldacena. *"If we lived in such a universe we would be, in some sense, approximate descriptions. But that's nothing new in physics. Take the surface of a lake, for example. It seems to be a well-defined surface; insects can walk*

on it. But if you look with a sufficiently powerful microscope, you'll see that there are molecules moving around and there is no sharply defined surface. The idea is that spacetime could be similar. It's not well-defined in an absolute sense, but we are so big that we don't notice it." Just like the insects on the lake, we'd be looking at the world with eyes that are too crude to reveal the true nature of spacetime. Ignorance is bliss, so in an everyday practical sense, whether or not we live in a hologram probably doesn't really matter — though there's endless fun to be had with the philosophical side of things.[70]

Brian Greene

According to physicist Lee Smolin, "there is a strong feeling among those of us in the field that some version of the holographic principle is true, and if it is true, it will be the first principle which makes sense only in the context of a quantum theory of gravity."[71] Holographic theory is a physical theory of processes where matter, spacetime, and information are *dependent arising* inputs and outputs of computational processes. What process relational philosophy and the physical theory of a holographic universe suggests is a computational universe, one where the processes or algorithms are the universe's fundamental building block.

It is important to interject here on a common fallacy of metaphysical hypothesis and explain how we are not taking the bait. The meme of a technological breakthrough spread for years or even millennia, and no institution of man is free from the gravity of the societal change that new technology can bring—not even metaphysics. Lo, metaphysics has had a terrible track record of associating existence with the latest technology using geometry, clockwork, and even briefly, a steam-based explanation as models for cosmic function.

A computational universe, like the one proposed by process philosophers is not taking this well-trodden, dead-end path for one reason: a computer is a universal machine, or, more accurately, a computation is a universal process. Alan Turing proved that computation could be done on information independent of the substrate—be it neurons or silicon or nodes on a web of spacetime—so long as sufficient memory storage and speed was available to finalize all of the instructions in the computation. Computation is not a technology as such, but a fundamental process, like entropy or fields.

Our universe is operating on processes: projections of information, quantum computations related intrinsically to the geometry of spacetime, and, most importantly for the thesis of this book, optimizations—processes that have direction, a purpose.

The holographic universe gives us our first glimpse at a physical system that aligns to an important component of the metaphysical framework of process philosophy—the idea of an *object* universe that has encoded in it all of the information that is compiled and projected to make our own universe.

The holographic universe is only one sort of computational universe. David Deutsch has already begun in the construction of the underlying laws that would govern the translation of information into physical entities—like in a hologram—or any other computation allowed by the laws of physics. Deutsch's recent project, in partnership with Chiara Marletto, is called constructor theory and aims to develop a library of *constructors*—where a constructor is a physical entity which is able to carry out a given task repeatedly. A task is only possible if a constructor capable of carrying it out exists, otherwise it is impossible. Eventually, a computational theory of the universe could be fully explained through an understanding of the interaction of different constructor subroutines. In the next section, we will discuss how a pure computational universe arises from our development of virtual reality and what our experience with these constructors tells us about the possibility that we are living in a simulation of some previous "real" existence and the likelihood that we can develop a similar virtual experience for our machines—including conscious artificial intelligence.

The Red Pill

In order to understand the algorithmic nature of the universe, consider virtual reality. The *virtual-world-in-itself* can be encoded to be a fantastical world or something that represents reality with great fidelity, the latter of which we will consider. Think about how this virtual reality is created. The universal classical computer[72] hooked up to a headset runs through a number of parallel processes encoded in some computer language. Focusing on the algorithms in this computer language, we will see many equations that we recognize from physics: $F = ma$; $n = c/v$. The more detailed these equations in the rendering of the virtual world, the less virtual and the more realistic the world becomes, to the limit proposed by David Deutsch that universalizes virtual reality in the same way as Turing universalized the computer. Universal virtual reality constructors can be used to create simulated universes with perfect fidelity to our universe. Our universe must be computational if the processes used to create more and more precise virtual worlds are the exact ones that we measure in our real world. Just like in the virtual world, existence is rendered from algorithms in a universal computer.

> *It is possible to build a virtual-reality generator whose repertoire includes every physically possible environment. So a single, buildable physical object can mimic all the behaviors and responses of any other physically possible object or process. This is what makes reality comprehensible.*
> David Deutsch

This is an amazing leap forward in our understanding—a simplification of existence! Our universe being computational clears up not only the mechanisms of the most paradoxical physical processes—like entanglement—but also describes why existence can be explained at all. One of the strangest but altogether necessary properties of existence is its comprehensibility—the fact that we can figure out how it works. When you really think about it, this is a fantastic property and one that we take for granted. When you write out an equation like, $H\Psi(x) = E\Psi(x)$, the Schrödinger equation, you are not just describing a process, you are decoding the source code by which all of existence works. It doesn't have to be this way; in fact, it would not be this way if the universe did not arise from a system of computations. The laws of physics, more aptly named the algorithms of nature, must be the root to the generation of reality since we can work backwards and find indications that they are responsible for the way nature works.

If we wanted to only build a simulation, a high-fidelity facsimile of existence, we would stop with our classical universal virtual reality generator. Our universal computation would present into our world a near perfect simulation of our reality, but it would not generate a reality, it would require our current existence and experiences to perform its matching trick. It would also be extraordinarily slow and would not have memory enough to simulate even a modestly complex universe. We have one upgrade still to make in our computational universe VR generator before it has the power of generating a comprehensible universe like the one we live in: an upgrade to the type of computations it is running and the type of information it uses in its processes, an upgrade to a quantum computer.

> *For Deutsch, a universal computer had become nothing less than the key to understanding reality. Such a machine, being able to generate all physically possible worlds, would be the consummation of physical knowledge. It would be a single, buildable physical object that could describe or mimic with perfect accuracy any part of the quantum multiverse.*[73]
>
> Jim Holt

The use of a quantum computer brings process ontology full circle back to materialism and idealism. In a universe that is fundamentally a quantum computer, like with the holographic universe, the process, material, and informational worlds are completely described and align to our latest explanations for existence. A quantum reality generator uses the algorithms of nature—especially the wavefunction algorithm—to create a superposition of states in information storage which are available to the collapse and generation of material reality. We do not need to supplement our explanations with newfangled probability waves to describe an algorithm storing probabilities into information centers. Furthermore, like the progression of science from classical to quantum, as we reframe the universe as computational, we utilize modern conceptions. The standard model and relativity work exactly as they always have, they are just framed as the fundamental algorithms that *compile* our reality, rather than abstract equations that have no significance beyond their ability to describe existence.

Existence is fundamentally a quantum computer. This frame for the nature of reality should follow logically from what we have learned about the world. We have changed conceptions about the nature of existence before, but this restatement gives credit where credit is due, it states what the entirety of the physics community either boldly proclaims or keeps hidden in subtext when referring to the laws of physics—that the processes we have written in mathematical language and observe time-and-again—are running the show. Everything from their accuracy to their incongruities to the fact that we can comprehend these universal mechanisms is due to the algorithms of nature existing as algorithms, processes that serve as the instructions to run everything from the dynamism of particle interactions to the slow dance of gravitational attraction. An existence fundamentally processing through read-write-learn-compile, utilizing information and matter, enables more than just a description of the most basic building blocks of existence, but also the nature of knowledge and even our conscious experience in a computational existence.

> *So, the laws of physics not only permit (or, as I have argued, require) the existence of life and thought, they require them to be, in some appropriate sense, efficient.*[74]
>
> David Deutsch

The algorithms of nature are capable of meaningful learning. We don't have to imagine what these learning algorithms look like; we have examples of some of these "smart" subroutines in processes closer to home: selection process pressures from genetics and the development of knowledge in epistemology. In his book *The Fabric of Reality*, David Deutsch unequivocally argues that since complex lifeforms exist, there is a computational subroutine—in his words a constructor—that we know as selection, which gives rise to complexity in the face of entropy through a process of randomization and error correction to (in the case of life) environmental pressures. Deutsch references the work of ethologist Richard Dawkins on the propagation of memes as another example of the computational universes' use of the selection constructor to select for the most popular idea.

Deutsch goes well beyond Dawkins in his follow-up book, *The Beginning of Infinity,* on the epistemological subroutine in a computational universe[75]. The epistemological process of creating better explanations—which Deutsch defines as knowledge of the nature of existence—coupled with a universal constructor that can utilize resources to make things, is only limited by the laws of physics. A virtuous cycle of problem solving to create the next machine to tell us more about the nature of the universe to create the next machine is a motivation—a purpose in science and technology. The computational universe outsources the writing of new learning—through our knowledge—into the universe. In just our little corner of the universe, we have figured out a way to cool some small portion of a vacuum down to one billionth of a degree above absolute zero (-273.15C), the coldest temperature in the known universe. We know that if we find a lower temperature in the universe, closer to or achieving absolute zero, it will be due to knowledge, not purely physical systems. By framing these acts as subroutines in a computational universe, we are again thrust into the center of a meaning greater than ourselves, where our collective wit in writing the solutions to the questions read to us by the universe is a deep learning problem writ large.

Like the materialist conception of poetic naturalism where we are "the eyes of the world," the universe's experiential mind's eye, the computational universe using us as an epistemological deep learning constructor gives us another sort of meaning in the multiverse. There are only two things that can manipulate the course of the universe, the algorithms of nature and knowledge. Across the great expanse of spacetime that we have studied with our intricate telescopes, taking in every radiative wavelength, we see the march of entropy as unabated... except, in very local, highly engineered systems, right here on earth. We have extracted and refined energy from the planet and from our star to locally counteract disorder, performing useful work and resurrecting universal machines from the *terra firma* of what is possible in existence. There are likely other species in the universe with the ability to generate knowledge and its spinoff technologies, maybe some that started with richer resources in their interstellar space or had fewer sociological disruptions or have just had more time, but this makes our utilization (if that is what it is) by the universe in the development of explanations, technology, and knowledge no less special. The "royal we" collection of species capable of writing knowledge into the computational universe may be a more important part of the cosmic plan than we imagine. Our learning might be the sole input device for the computational universe, our ability to solve first the fundamental equations and eventually the construction of larger energy resource projects might enable a computational universe to turn expansion into a contraction and bring about a reboot—a big bounce—or any number of science fiction scenarios. If our explanations grow in progression with our ability to resource their spinoff technologies—like is hypothesized by the Kardashev civilization scale where a class II civilization is able to mine the energy of solar systems and a class III civilization is able mine the energy of galaxies—our deep descendants may have the computational power and understanding to manipulate the symmetries thought to be responsible for universal expansion or compression and write this back to the computational universe. The expansion of knowledge is only bounded by the laws of physics and we stand at "the beginning of infinity" in relation to its possible manipulations of the universe.

Intelligence like ours (and far greater than ours) carries out the deep learning algorithms and construction of novel complex technology in synergy with the aims of the computational universe. This meaning may be the most substantial one yet proposed by this book. There is profound potential suggested by the impressive subroutines of knowledge we have already been able to write and compile to manipulate existence. We have the knowledge to recreate vacuums comparable to anywhere in space; to create lower temperatures than anywhere else in the known physical universe; to split atoms and fuse them together; to record the interactions after the collision of hadrons after racing them to speeds nearing that of light; to "land a man on the Moon and return him safely to the Earth"; and to create universal classical and quantum computers. And all of these in the last fifty years. If humanity had social solutions to match our scientific solutions, we might well live up to our meaning for the multiverse.

PROCESS ONTOLOGY

It may be that our knowledge and ability to create complex technology to overcome entropy is what the universe needs from us. Science in the far distant future might arrive at a dialectic between explanation and entropy, a future civilization capable of manipulating the computational universe that might reverse the course of the heat death through a few manipulations of symmetries in the fundamental constants of the standard model. Our relationship to these deep descendants is certainly genetically tenuous, but if the survival of knowledge is essential to the reboot of the material universe, than we must continue to solve the existential problems like nuclear proliferation, climate change, global pandemics, and superintelligent AI if we are to have any chance of becoming such a heroic civilization... a civilization that is not "the eyes of the world" but instead "the engineers of the world."

Fig. 18. - **Engineering of the World Meaning** - Our ability to imagine and build complexity to overcome problems is a novel trait in the universe. Intelligent entities' (like humans) engineering prowess is the result of the multiverse's distributed deep learning algorithm running either out of curiosity, or, as a progression toward our deep descendants who might one day be called upon to solve a dialectic question of multiversal importance.

There are many possible local optimums to be selected by a purpose-driven computational universe equipped with deep-learning capabilities, resources, and constructive abilities. Independent of the need to reach a heroic and universe-altering technology sometime in the far-flung future, intelligent life's ability to create universal constructors (Deutsch proposes that universal constructors are only limited by the laws of physics) offers a plethora of possible technological futures. One of the most desired uses of our infinite progress in knowledge must be the simulation of the embodied, felt experience most associated with our subjective consciousness.

Unlike the information of other types of cognition now spread virally through memes, our ability to transfer information of our subjective states to one another—or more poignantly for this book as we move forward—to computers and the computational universe, is limited to esoteric approximations. Can we put experience into a computational existence in such a way that it propels advances in an explanation for consciousness? This will be one of the grandest undertakings of this book, a path forward for understanding experience through existence. Our first steps require us to take an instructive step back into the VR that compellingly argues for the computational nature of existence and deconstruct the problem on hand for understanding what might be thought of as the *experience subroutine* of our computational universe.

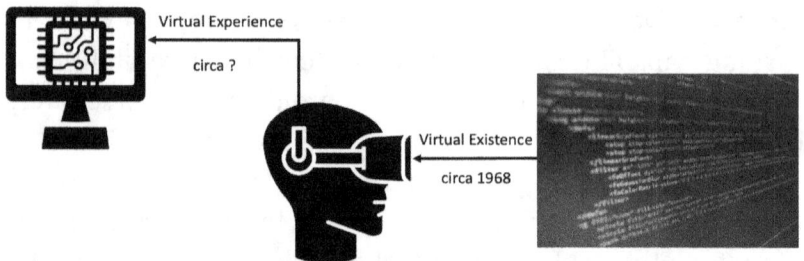

Fig. 19 - Virtual Existence is already available to us and the better the fidelity of VR's code to the laws of physics, the better the fidelity to reality. What leaps are needed to offer virtual experience in a feedback loop to classical computers?

In Figure 19, we recall that a virtual existence can be generated with perfect fidelity to our actual existence given our adherence to the laws of physics in coding the VR. The uncanny valley of visual VR to visual reality has already been crossed at the time of this writing. You can go to your local Best Buy and put on a VR headset and notice autumn leaves falling or (far worse) your grip on a rock wall failing, in real enough ways to cause your brain to treat it as real and, in the case of the examples above, cause you calm or calamity. But, the line out of the top of the head and into the classical computer labeled *virtual experience* in Figure 19, is still utterly mysterious. It is a microcosm of the hard problem of consciousness. From now on I will separate this microcosm out and refer to it as *the hard problem of virtualizing consciousness*. As we now understand from Deutsch's arguments for the universality of virtual reality, the process of learning how to virtualize something is an act of computational design that starts with a list of requirements. Our march from virtual existence to the fundamental nature of a computational existence started with first, the requirements of algorithms coded in the virtualization, and second a constructor like quantum mechanics to create more than a simulation. The same requirements— a set including both algorithms and a constructor—are sought for experience.

As we have discussed, as a matter of experience, there is only consciousness (embodied, first-person, felt experience) and its contents (sensory phenomena and thoughts). For the purpose of determining whether our experience could be replicated in virtual reality in the same way as existence is, let us break down each component individually and see what we can learn. Our designed, thought experiments will attempt to individually toggle the key factors of experience (sensory input, consciousness, and thought) and look for a response in the generation of a high-fidelity *virtual experience.*

First, we look at the overloading of sensory experience in modern VR. Sights, sounds, and even our sense of touch can be amplified in this virtual world. When programmed with high fidelity, we can be made to see an earthrise, hear the sound of the jungle waking up, or feel our ever-tightening grip on our rock-climbing handhold. In such extreme situations, our consciousness reacts as if the reality was not virtual. When put into a high-fidelity virtual reality—such as walking a plank set between high-rise buildings—most individuals with modest-to-extreme fear of heights will not be able to overcome their conscious commitment to the danger of the situation. Modern virtual reality generators like the Oculus or Vive greatly impact both the content of sensory experience and the construct of felt experience with improvements in full-motion, body-sensor virtual experiences like those being developed at The Void near Salt Lake City, Utah.

One of the most interesting developments that could be tested in VR and uploaded into our burgeoning virtual experience is what is known as the interface model of perception simulated in the laboratory of Donald Hoffman. The interface model of perception holds that our view on reality is not how true objective reality is but instead a user interface that we create around potential fitness outcomes. Hoffman has simulated that under natural selection pressures, genetic information is more likely to propagate if a species ignores truth and instead constructs its reality around fitness functions[76]. The user interface that humanity has created is itself a virtual reality that promotes evolution and subjective interaction over existence's essence. Using the simple algorithm of "fitness beats truth" in our virtual experience might at once simplify perception modeling and make it more alive to subjective feelings and narratives.

Another thought experiment useful in envisioning a path to a robust virtual experience involves this time controlling for sensory input by utilizing a sensory deprivation tank to highlight thoughts and consciousness. In this subroutine of our *virtual experience,* we are blindfolded and wear noise canceling headphones, our sense of touch is brought to equilibrium temperature and buoyancy through a body-temperature salt-water bath. Smells are generally neutralized, and tastes can be dulled by normal hygiene. We are left alone with our thoughts in the construct of consciousness.

The experience of our mind that is easily repeatable in virtual experience is the never-ending flow of thoughts as snippets of text, bits of internal monologue, and fuzzy images. Most of us think this constant brainstorm of thought is a bug we'd rather be without especially in a sensory deprivation tank where we are left alone with our thoughts. Surely, we'd immediately eliminate it in virtual experience. While the humdrum of thought can be a valuable tool for understanding how mindlessly most thoughts arise, it is also the greatest distraction to discover the nature of the component of experience hardest to virtualize—consciousness.

A titration to pure consciousness—the context where all subjective experience, sensory data, and thoughts act as content—is catalyzed by the addition of the society's current neuropharmacological offerings, especially psychedelics like DMT or psilocybin. Depending on numerous factors including chemical, dosage, setting, and susceptibility, psychedelics can offer alterations to consciousness like the complete disillusion of the sense of self, a feeling of oneness with existence, a gracious lovingkindness, and a limitless world of thoughts—both positive and malevolent. While there is still much discussion on the appropriate use of these chemicals in the treatment of illness and in the ritual expansion of experience for the well, there is no question that they offer a version of virtual experiential reality that is profoundly altered both during the "trip" and long afterward—often changing individuals' ideals and sense of meaning.

The durable alteration of consciousness with psychedelics offers some fundamental attributes for use in the virtualization of experience. Selflessness is a primary attribute that expands into a oneness with a conscious idealistic field during some psychedelic experiences. Timelessness is prevalent in both the meditative and medicinal experiments on consciousness. Consciousness during and long-after a "trip" is found to be meaningful with most participants ranking these experiences as some of the most important and instructive of their whole lives, bringing about changes in their further investigation of the nature of the mind, loving-kindness toward others, and a more ethical and enlightened stance toward nature and our place in it[77].

We are able to model existence through high fidelity VR technology but virtual experience remains a technological mystery largely because we know more of the algorithms for physics than for neuroscience. However, we have begun to experiment on ourselves, controlling for all but one factor in experience, in an effort to understand these processes more fundamentally. In the near term, we can imagine sensory deprivation, guided, psychedelic VR meditations being used to control for sensory input, reduce feelings of the self, and lengthening periods between being mindlessly lost in thought. Gamifying the objects and levels of focus, the use of interactivity with scientifically-controlled sensory components like slight changes to temperature or ambient noise, and body and brain monitoring will not only enable feedback loops for users to make improvements to their subjective states, but also begin to drive solid neurophysiological models about how meaningful states of consciousness can be achieved and made more persistent.

The algorithms for use in coding virtual experience are unknown. There are numerous reasons why consciousness is not well-described in simple mathematical language, but I do not believe that it is because the problem of finding these algorithms is without a solution. There is no equivalent Schrödinger equation for experience. However, the constructor is very likely the same, and we should be looking for a *wavefunction of the mind,* an explanation for how experience arises not as a result of classical physics and computation but utilizing quantum mechanisms. In the next chapter, we will see how the *wavefunction of the mind* might be solved for by using the parallelism inherent in the many world's interpretation of quantum mechanics. The use of a quantum computer as the constructor, to receive the algorithmic wavefunction of the mind might just be able to solve the hard problem of virtualizing consciousness as shown in Figure 20.

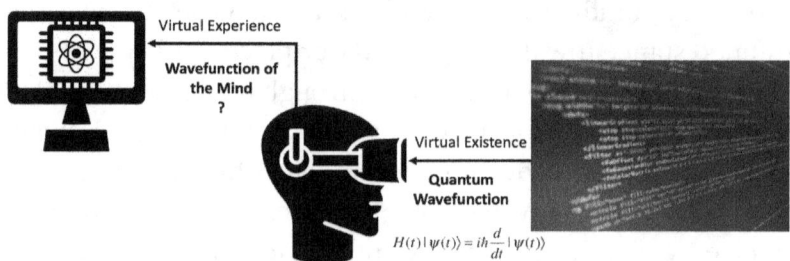

Fig. 20 - The hard problem of virtualizing consciousness is made more difficult by the lack of an algorithm and a computational platform powerful enough. As quantum computation continues to make technological gains, the wavefunction of the mind algorithm is more likely to become solvable.

Read-Write-Optimize

The universe—its beginnings, existence, mechanics, and our relationship to it—is well explained when process is the fundamental constituent. Processes and their relationships are responsible for it and bit, space and time, fields and forces, causality and existence. Some of these algorithms have been discovered, plucked from the source code known as the laws of physics, while others remain obscured. Since the natural world is fundamentally algorithmic, there are imperfections, incompleteness, local optimums, randomness, complexity, and even intractability. Computational limitations must be considered alongside accuracy, precision, capacity, and speed.

Processes can have purpose. Like a recipe followed by a cook, not only does a meal result, but paired with the process of presentation and the economic process of pricing, an optimum can be attained. Algorithms are designed to step through problems, creatively but imperfectly using logic and chance, trading accuracy for efficiency. Algorithms can be designed to the purpose of defining a relationship's equilibrium (game theory), predicting the future (Bayesian inference), or finding the best solution with the least computational energy (optimization).

A universal optimization algorithm could be a process that is persuasive but not all-powerful. This optimization process would be related to all other processes, those we know and reverently call the laws of physics and those we don't yet know, and is one template for universal progression toward a purpose.

Optimums, and indeed all computations, are never perfect. The universal optimization algorithm would not employ supernatural subroutines but instead would do what makes the most sense in the least amount of time. Up against such hard cases, effective algorithms make assumptions, show a bias toward simpler solutions, trade off the costs of error against the costs of delay, and take chances.[78] Like algorithms in computer science, the universal optimization algorithm could call upon both abstraction and on the discrete, on approximation and precision, probability and determinism.

That our universe is fundamentally a system of related processes—some already partially uncovered, explanatory, and therefore utilized, while others remain obscured to our modern inquires—is a plausible hypothesis. A computational universe, fundamentally comprised of processes, could have a true optimum: simply an algorithm with an objective function improving on other local possibilities. Our experiences could be more than simply poetic or profound, they could be in a read-write relationship with the various routines in the computational universe, including the persuasive but passive optimizing algorithm. The hypothesis that these universal processes correlate to something meaningful for human existence *and* that the processes of our experience can relate and interact with the processes that construct existence is speculative. The fundamental nature of process in our universe offers a good reason to believe that optimization processes might be available to our universe and to savvy intelligent beings. However, the fact remains that our universe does not appear, at present nor at any time back to the big bang, to be optimizing.

Instead, the physical world behaves in accordance with the processes that are consistent as far back as we can see, entropy and quantum mechanics. Our distant ancestors will likely live in a lonely place where intergalactic space is stretched so wide that all of the light we now see will be far over the horizon, but with the same roiling probabilistic evolution of fundamental components processing per the wavefunction.

However unlikely it is that we will observe our universe racing to line up along some celestial or subatomic optimum, a focus on our universe, as odd as it may seem, is likely a limited and parochial myopia. Far from being fringe physics, the idea that a multiverse exists for each quantum distinction has extraordinary explanatory power and may be the simplest description for such strange behaviors as observed in the dual-slit experiment and for the physically impossible but unshakable subjective sense that time flows. In the next chapter, we return full circle to the multiverse and will discover that in the set of scenarios available to many-worlds existence, a purpose-driven and even participatory process is still possible.

CHAPTER GLOSSARY

process relational ontology - claims that processes (algorithms, wavefunctions evolution, optimization, consciousness) and their relationships with things and information are the fundamental constituents of the universe.

Tao - roughly translated as "The Way," it is the first known process relational ontology.

holographic universe theory - a computational universe that compiles quantum information on a lower-dimensional "surface" into the holographic reality in the higher-dimensional "bulk." This process of projection has been used by physicists to calculate key quantum gravity interactions previously intractable.

universal virtual reality constructors - can be used to create simulated universes with perfect fidelity to our universe. Our universe must be computational if the processes utilized to create more and more precise virtual worlds are the exact ones that we measure in our real world.

universal optimization algorithm - a process in a computational universe theoretically able to bring about universal meaning.

KEY TAKEAWAYS
- Process relational ontology is suggested in the ancient metaphysics of Taoism with its focus on an underlying process—the Way or the Tao. All information and material are thought to be emergent from the Tao.
- Both relationships and processes figure heavily into theoretical physics' most promising theories of quantum gravity. Algorithms may form the laws of physics and utilize the process of projecting our holographic universe from the information stored at the surface. This process of projection is both read and write.

CHAPTER 6:

Many Worlds

CONCEPTS
- The many worlds interpretation of quantum mechanics is explored and shown to be the best explanation for: the interference seen in a single-particle dual-slit experiment, the experience of the flow of time, and the computational edge of quantum computers over classical computers.
- Meaning stemming from the parallelism of the universal optimization algorithm and of the interaction of experience across *near-parallel* universes is explored.
- We investigate how consciousness (and, thus, moral alignment) might be engineered into our machines and what paradoxes arise because this has already happened in the multiverse.

DOES THE REFRAMING OF the world as processes make sense? That is, does it adhere to our scientific and rational formalism? This *edge-on* view of science calls on the laws of physics to act in the leading role and gives us an interesting perspective to approach the question of meaning. However, as much as we look for the proof of an ontological optimum in the universe, we see little indication of it. Particles don't line up so we can count their energetic vectors, but instead we have an emergent property like temperature that gives us their statistical dynamics while the standard model describes their interactions. The law that explains our universe—at least at the most fundamental regimes and all the way back to just moments after the big bang—is quantum mechanics. The process at the core of the standard model, the wavefunction of the universe, does not appear to be any sort of optimum. Where true panpsychic idealism—the idea that even quantum particles are conscious in some sense—could be obscured to us just as the subjective experience of other living creatures is obscured to us, it is not the case that a universal optimum ontology would go unobserved by our instruments or mathematics.

However, there is one remaining wrinkle that we need to explore before dismissing our quantum mechanical world as one that is not manipulated by an optimization algorithm and that wrinkle stems from the same sort of interference that keeps us from observing a superposition of dead-alive cats in Schrödinger's paradox—the interference from near-parallel universes in the multiverse.

Dual-Slit Experiment

Take a wave of water and put it through two parallel holes and measure the interference as the newly distributed wave crashes against the far shore of the measurement device. The measurement device will show an interference pattern where troughs and crests are distributed according to the interference pattern of the wave in question as shown in the figure below.

Meaning in the Multiverse

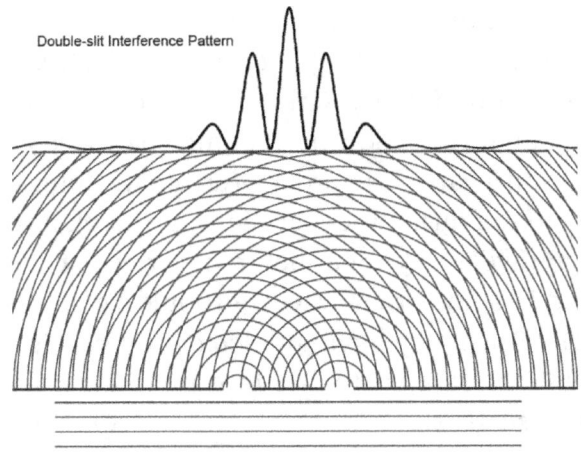

Fig. 21 - Dual-slit Interference Pattern of a Wave

If instead a droplet of water is fired through the slit on the left, a single measure of moisture is found on the shore at the far side aligned behind the left slit; while the same measure of a droplet's worth of moisture is measured behind the right slit if the droplet is fired through the right slit. Finally, if one's aim is off with the droplet water-cannon and the droplet is fired into the wall in front, no measurement is made on the back surface.

Shinning a light through the dual slit produces an interference pattern similar to the wave of water. Knowing that the beam of light is comprised of photons in a way not altogether dissimilar to how water is composed of molecules of H_2O, we hypothesize that photons interfere with one another. Where the light beam shines against the wall with the dual slit and beyond to the screen, there is interference, where there is no light beam, there is no indication of interference, as we would expect.

What is beyond our expectations is that when a single quantum particle, individual photons of light or hadrons of matter, are fired at the dual slit screen, the interference pattern of a

MANY WORLDS

"wave" returns! In other words, unlike the single droplet of water, a single quantum of light—in the case of a photon, a massless, point particle—is creating interference with itself!

But there is one further kick to our intuitions that further flummoxes our formulation of quantum particles as acting as both a wave and a particle in order to create the interference pattern we see in the dual-slit experiment. If the photon was truly acting like a wave at any point before it arrived at the screen or was splitting in some other way, we should be able to measure this property. Setting up a photon detector at each slit shows us that this is not the case, that indeed we are tracking a single photon through a single slit. The interference is consistent with our changes to the number and timing of slits that we open, that is, the interference with a single photon or hadron with two or four slits open is the same as that for a light beam of many photons with the same number of slits open. There appears to be interference—with other photons—in both the case where we supplied the other photons in our beam of light, *and* in the instance when we only supplied a single photon.

The problem of the appearance of single-photon interference led Hugh Everett III to hypothesize that there were other photons—in other parallel universes—interfering with the single photon in our universe in our dual slit experiment. For every quantum particle, there are hosts of *shadow-particles* that are only indicated by their interaction—an interference in perfect accord to quantum mechanics—with the *tangible-particles* in our universe. (And our tangible-particles are responsible for interference, playing their role as shadow-particles, in the parallel universes.) These shadow-photons occupy other positions calculated to be possible by the wavefunction equation of the quanta, which happen to be their actual positions in *their* universes, and the only indication we have of their existence, and the existence of their parallel universes, is the interference they cause in our universe.[79]

> *Single-particle interference experiments such as I have been describing show us that the multiverse exists and that it contains many counterparts of each particle in the tangible universe... The heart of the argument is that single-particle interference phenomena unequivocally rule out the possibility that the tangible universe around us is all that exists.*[80]
>
> David Deutsch

The simple single-particle dual-slit experiment confounds all but one explanation of the quantum world, the many worlds interpretation. Individual photons fired at a scientifically-sliced barrier interfere with trillions of *shadow-photons* and create an interference pattern on the screen comprised of atoms, while trillions of parallel universes record (on screens made of *shadow-atoms*) an ever so slight deviation in the interference of their photon with the trillions of other *shadow-photons* and "our" single object photon.

Unlike most other theories of quantum mechanics, the multiverse or many worlds interpretation of quantum mechanics explains what is observed in the dual-slit experiment. The mathematics of the wavefunction of a single quantum evolve more naturally in a multiverse, collapsing according to the probability of being interfered with by the parallel quanta in parallel universes. The wavefunction evolves in parallel universes in exactly the same way it does in this one, the probability of collapse at a particular place being dependent on the shared-history of interference with other particles, be they quantum particles of "this world," observers, or the same entities in other universes in the multiverse. In strict adherence to the Schrödinger equation, the many-worlds interpretation of quantum mechanics frees the wavefunction to act continuously across the entire universe—the wavefunction of the universe—with any alterations being not to its makeup, but to the system of near-parallel universes and their universal wavefunctions. It is parallel processing of the wavefunction algorithm with the quantum multiverse as the construct.

> *I've found that all too often, people who learn about, or are even somewhat familiar with, the Many Worlds idea have the impression that it emerged from speculation of the most extravagant sort. But nothing could be further from the truth. The Many Worlds approach is, in some ways, the most conservative framework for defining quantum physics.*[81]
>
> Brian Greene

While there are difficulties in arriving at empirical evidence for the many worlds interpretation of quantum mechanics, it is not impossible that evidence (greater than the distinctly quantum interference pattern in single-particle, dual-slit experimentation) will be found. There are subtle differences in the predictions of the multiverse and the Copenhagen interpretation that could be used to experimentally resolve whether the many worlds interpretation is the correct formulation.

Adherents of the many worlds interpretation of quantum mechanics like David Deutsch believe that parallel universes are real and related to one another—at the very least in the interference they create between quantum fields and particles. The dispersed interference pattern of a single particle passing through a dual slit is caused by the *shadow particles* that interfere *as if* they were photons... *are* photons. These shadow particles *are* particles, just ones that exist in near-parallel universes.

It is important at this juncture to describe the multiverse. If you are like most people, you imagine the nearly infinite parallel copies of you, some that are nearly exactly like you down to the quantum level, while others are your evil (or nice) parallel. Indeed, this is part of the picture. What I have termed *near-parallel* universes are those that most closely resemble our universe, interfere in our dual-slit experiments, and include vast distributions of doppelgängers nearly indistinguishable from you. By contrast, a *far-parallel* universe does not contribute information to our quantum computations, does not interfere with our wavefunction or the experiential flow of time, and has evolved to be very distinct from the instance we will consider "our universe." We will return to the importance of far-parallel universes later in this chapter.

Any single universe is frozen in space and time, with any dynamism coming from the interaction of the *kernal program*—the wavefunction—across fungible near-parallel universes. In the many worlds interpretation of quantum mechanics, the wavefunction is not probabilistic across near-parallel universes, it only appears so when restricted to our view of "this" universe. Where we believe we observe a collapse of a local quantum event into a distinct particle, the wavefunction of near-parallel universes is in an interference pattern with "our" world and each of these other near-parallel universes' wavefunctions take on the other probable outcomes.

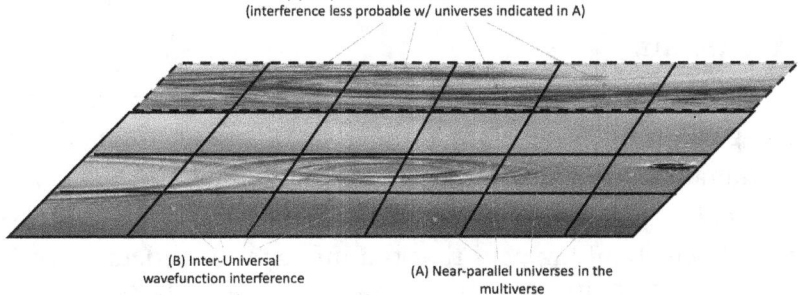

Fig. 22 - *(B)* Interference from the evolving wavefunctions of near-parallel universes (A) simplifies the standard model to its most consistent explanation. Far-parallel universes (C) are not fungible with the universes at (A) and do not appreciably interfere.

Conceptualizing the multiverse as a parallel processing computer is useful at this stage. The computational multiverse has many parallel-processing programs running in ways similiar to the wavefunction. Some of these are known to us, the laws of physics, some we are yet to discover. Since the multiverse is a quantum computer, it projects a material universe and not a high-fidelity virtual reality. The flow of time, projection of the material universe, and even our conscious experiences are all emergent from these parallel processes. The difficulties of describing *apparent* frozen continuums, collapsed states, and conscious hard problems arise from our inability to perceive "outside" our universe or conceptualize a computer powerful enough to run all of reality & our experience of it. The mechanism by which the multiverse works is the parallel processing of programs on a multiversal quantum computer.

For our purposes, since the multiverse offers one of the only explanations for the evolution of the wavefunction while maintaining all of the theory's predictive power, we will consider not only the implications of this massive parallelism on the material in the multiverse, but also the implications of the interactions between parallel universes on the informational components, the processes of computation, and our experience.

Parallelism

Our universe is part of a multiverse where the wavefunctions of near-parallel universes interact. However, our mathematics and the metaphysics suggest that more of existence is available to interact in some unique way in the massively parallel multiverse. Indeed, we should expect that the informational data stored on the surfaces of all near-parallel universes should offer the near continuous capacity needed to store a highly entangled system of quantum particles or, more speculatively, store high fidelity remixes of our lives in their entirety. Furthermore, it is logically consistent that there are other processes we have yet to discover that are optimized for parallel processing. Our search for meaning no longer has to trounce only through our visible universe, but can also look across the stack of those universes that are near-parallel to ours, distinguished from ours by just a few quantum events, but in constant interaction with our existence and, as we have already seen with the flow of time, our experiences.

In our determined universe where space and time are combined, time can only *flow* as the interaction of near-parallel universes that have a slightly increased entropy according to their universal wavefunctions. The dynamic interference of near-parallel existence for the experience of time's flow, gives us some indication of the immensity of even the near-parallel multiverse, and makes any reference to "our" universe, problematically parochial. The number of universes *just* interfering to cause time to flow along its entropic arrow is beyond astronomical since each quantum distinction is an indication of a modest evolution of the wavefunction of a near-parallel universe that pushes time forward another diminishingly small unit. Our universe in the multiverse is only ours very briefly but is infinitely parallel.[82]

Having learned from our wanderings out of the paradoxical *frozen flow* of time into the light of interference between near-parallel universes, we should remain open that our conscious experiences are not only a part of existence, but can also tell us something about it. Our conscious experience of the existence of both space and time opens us onto universal meaning. Reframing our relationship with *the continuum*—that at the fabric of space we are interrelated to everything and that our experience of the dynamism of interference in the multiverse is what we call time—is the kindling that smoked out the relationships of conscious experience to the reality of universal existence. Going forward, we will tie our materialistic, idealistic, computational universal meanings together even with our most ardent personal meanings by continuing to find explanations by looking not down just at "our" universe, but instead across numerous parallel universes.

Computational parallelism, where the execution of processes are carried out simultaneously, is a key feature of the speed and dexterity of modern computers. As much as possible, instructions are broken down into subcomponents and executed in parallel to utilize the resources of multicore architecture. Instead of relying on faster hardware to speed up the computational frequency, performance enhancements now more readily come from an algorithm's improvements in parallelism.

There are three types of multiversal meanings to consider, the first of which takes up where we left off in the last chapter—universal optimums in a process ontology that may not be noticeable in our universe because these routines run across parallel universes. The optimization algorithm's use of near-parallel universes to run experiments to select optimal outcomes offers a logical reason why we do not lead lives where constant signals from the universe compel us toward our best selves. Like the interference of near-parallel wavefunctions impacts the wavefunction under study in our world, so too might the multiverse's optimization routine run simulations both in our universe and other near-parallel universes and impact our own meaning. This distributed learning through parallel simulations is the way that the artificial intelligence computer program Alpha Go learned to play Go, Chess, and Jeopardy and is the same algorithm we use when we hope to optimize our experiences through visualization.

Visualization is just a mental model, a thought-experiment where you place an avatar of yourself into the situation of peak performance. The more highly detailed the visualization, complete with a move-by-move account of your perfect routine, the more likely the visualization will help your performance improve. In the near future, this visualization will not depend on your imagination but could be beautifully rendered in a virtual reality. My old coach's saying, "you play like you practice," could be virtually true instead of a motivating hyperbole. Instead of a useful gimmick for optimizing performance, designed VR deliberate practice might enable real-world optimal performance to be more like deja vu.

Now imagine if, instead of only having the benefits of virtual reality to understand the consequences of various modifications to a designed routine, you had numerous, full parallel universes. Any parallel processing optimizing algorithms would have such a luxury and could utilize near-parallel universes to run concurrent distributed scenarios to understand the consequences and build local optimums from this machine learning in near-parallel universes.

But what would be the goal of these experiments? As we have seen throughout, the meaning might be a works project we participate in *for* the multiverse's ultimate meaning or as a form of grace from existence onto our meaningfulness *in* the multiverse, or both. The optimization algorithm falls into an opportunity for both works and grace purposefulness. If the optimizations are thanks to the grace of the computational multiverse, then the ordering of experience, the promotion of virtuous cycles that result in more optimum outputs, is part of the very fabric of existence. The reason for this grace has either escaped our parochial view of our path through near-parallel universes or is due to some manipulation of the multiverse that we do not yet understand, but lends credence to the idea that the multiverse is conscious in some way, experiencing for itself the benefit of this grant of grace. We will return to this superconscious entity later in this chapter.

On the other hand, if the optimizing algorithm is the machine learning mechanism used across near-parallel universes to increase the efficiency of all processes, then the universal meaning it ingratiates experience with is simply a by-product of the sorts of redundancy, parallelism, and innovation that our classical machine learning algorithms perform on our MPPs (massively parallel processors). Any peak experience is just positive proof of the effectiveness of the optimization algorithm and is input into the next scrum where new learning is tested and if successful, implemented. Any meaning we experience is parochial; the computational multiverse is actually just optimizing the process of experience, but just as we mistake the interference of near-parallel universes as the flow of time, we incorrectly frame meaning as personal and not resulting from any mechanisms of existence. A hypothetical example of this machine learning algorithm might include a tweak of existence that enhances the flow states of the favorite musician of a critical climate scientist, thereby improving humanity's advances against the existential threat of climate change.

As part of a multiversal works project, optimization processes might be used to persuade the *engineering eyes of the world* to meaningful experience or to promote existence engineering in alignment with some meaningful path *for* the multiverse. As discussed in previous chapters, the multiverse might be experiencing through us for its own quest to meaning; the multiverse may also be using our knowledge and ability to construct complexity to engineer a dialectic project that requires a highly advanced intelligence. In either case, the parallel processing multiverse may utilize the optimization algorithm to incentivize progress along a path of greater well-being which is persuasive of species-continuation necessary to advance knowledge or experiential heights so that the multiverse can achieve its own manipulation or meaning.

This sort of persuasion from existence is highly benevolent and was one of the outcomes of Alfred North Whitehead's foundational work in process ontology *Process and Reality*. Multiversal persuasion in the pursuit of its meaning or dialectic shows both the constraints of the laws of physics that keep us firmly rooted in natural solutions and the incentive that experiential well-being and societal shared meaning is for intelligent and conscious entities. While Whitehead drew upon the baggage-laden name of "God" to recognize the parallel-processing quantum computer most likely responsible for such loving persuasion, the recognition of what such magnanimity could produce is spot on.

> *God makes available and draws us toward a range of possibilities that enable us to envision a world beyond the world already actualized... God's persuasive work in us enables us to perceive visions of truth, beauty, and goodness that inspire the metaphor of a spark of divinity within every person.*[83]
>
> C. Robert Mesle

The read-write-optimize abilities of the optimization algorithm running in a massively-parallel process multiverse enables a universal natural meaning that *could* explain how purpose arises from the natural world. The optimization algorithm running as a machine learning optimization on the parallel process of consciousness—tweaking some underperforming subroutines and maintaining one of many processes—is an extremely pragmatic form of computational universal meaning whose subtleties match what we observe. The addition of increases in meaning as a persuasion, either to incentivize our experience or engineering, is a logical continuation of the meaning introduced as first the *eyes of the world* and later added onto as the *engineering eyes of the world*. If we are being utilized by the multiverse—either for our consciousness or in the methodical alteration of existence by knowledge—this parallel computational process would likely use everything within its means to ensure our societal well-being to our deep descendants. Those things within its means are not the tools of the deities of old; omnipotence is restricted by the laws of physics, but experiential persuasion is not.

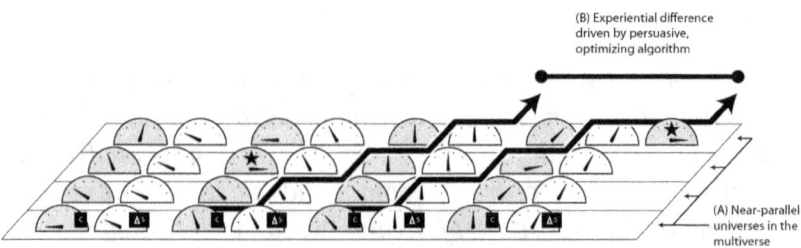

Fig. 23 - **Multiversal Persuasion & Optimization Algorithm Meaning** - Experiential interactions in near-parallel universes (left path) is persuaded toward slightly improved conscious outcomes (right path) by the universal optimization algorithm (B).

The *Persuasion and Optimization Algorithm* suggests a meaning where existence (the algorithm) manipulates experience (consciousness). However, this dependency still separates existence from experience, so it is time to take on this unnecessary dualism. Interference from a distribution of conscious states in near-parallel universes—the wavefunction of the mind—creates a new frame for how the hard problem can be virtualized and even overcome by taking a view across experiences in the near-parallel multiverse. Instead of a panpsychic conception of prevalent-personal meaning where the subcomponents are atoms and quanta, a *parallel*-psychic view where the components are near-parallel distributions of our conscious doppelgängers offers a more robust *prevalent-parallel personal meaning*. This idea, which I call the *many minds interpretation of consciousness* is an exciting view into the mystery of how experience arises from the parallelism of existence.

Consciousness from Clones

Scientific studies have modeled consciousness as an electro-chemical process, a quantum process, a holographic one, and a collective computational one. The true mechanism that gives you the rich experiential world is mysterious. Especially as we continue to develop greater cognition into our machines, we would like that they have the benefits of subjective experience, especially benefits like compassion.

Consciousness is a continuous and holistic process. As you know, your experience of reading or listening to this book now is bound with the feel of the fabric of your clothing, the smells around you, your awareness of my change in addressing you directly, and the background sounds that occur whenever we try to concentrate. Consciousness plays back our *first-person-close movies* from our memories and designs complete visualizations of our plans, seamlessly integrating past causes and hoped-for events into our awareness. Consciousness gives us the appearance of control and insight into the nature of our individual being. Furthermore, like all brain-based processes, including memory, learning, and language, there is every indication that we can improve our consciousness, developing awareness and insights of *what it is like to be* more interrelated to one another, to experience more fully, and to see existence as a process of becoming.

Yet our intuitions about consciousness, our explanations about our subjective state of being, are often flawed. One of the largest sources of persistent confusion is our identification with the *thought of self*. I want to point out the emphasis on the object of the last sentence—the thought of self—for that is what it is, only a thought, no more or less deserving of our awareness than our multiplication tables or a task list, and certainly not a thought that requires we identify with it. Whether you are aware of consciousness (mindful) or not (mindless) there is nothing in experience except consciousness and the contents of consciousness, of which, the thought of self is one such qualia or content. The subjective experience is not experienced by a separate observer or nexus of awareness.

What truly divides philosophers is not what consciousness does but how it does it. Is consciousness only brain-based or does it take up a *dual* residence in brains and something else in the universe? Do classical chemistries predominate or are the neural correlates of consciousness expressed through quantum mechanical means? Is consciousness transferable—either to another person, to a machine, into the cosmos, or to a heightened state?

Our conception of consciousness is solely thanks to the fact that we are conscious ourselves. We would not conceive of subjective experience if we did not have our lights turned on. There is no unconscious, objective measuring device for the degree of subjective conscious experience an entity is having. The Turing test was not meant to interrogate consciousness but instead intelligence. There is nothing you can hook up to brains or beehives to determine the degree of consciousness that is enjoyed by any of the myriad things.

How the system of consciousness emerges from the experiences of lower-level information processes is unknown. However, most philosophers of the mind believe that finding a neutral (non-physical and non-mental) process is a requirement for a scientific understanding of consciousness and a complete explanation of experience.

> *What will be the point of such a theory? If we could arrive at it, it would render transparent the relationship between mental and physical, not directly, but through transparency of their common relationships to something not merely either of them.*[84]
>
> Thomas Nagel

The mechanism I propose for resolving information processing at least at the level of human cognition into conscious states, the neutral interlocutor, is *experience interfering with other parallel copies of itself* in the multiverse. Experience as a distribution of many minds in *nearby* universes around the one experienced in this world, forms the *wavefunction of the mind*. The interference of the *shadow experiences* of a distribution of "you's" from unconscious-to-peak states resolves to the novel subjective experience you are having.

The solution to the hard problem of consciousness—that it is like something to be a collection of atoms—is the same as the solution to the dual slit experiment, that is, interference from near-parallel universes. Experience is dictated by a (yet unknown) parallel process: the wavefunction of the mind. Like our experience of the flow of time, our experience of consciousness is just our (understandably) parochial view of a slice of the myriad near-parallel universes interfering in each moment and in each qualia. Here we've reached a merger of conscious experience with the existence of the multiverse.

Much like a distribution of outcomes occurs in near-parallel universes on the screen of our dual-slit experiment, a distribution of consciousness arises in us and our clones in near-parallel universes. At one end of the spectrum, on the vast tails of the subjective-experience distribution, are unconscious zombies in far-parallel universes. Rising up from the tails of the distribution are our mostly mindless states, those unexamined states that make up the bulk of conscious experience where we are lost in thought, movies, phone games, or our emotions... the experience of most of our *shadow clones*, that sometimes entangle us in their grasp. The crest of pure consciousness is contrasted against the experience of a clone dampened in their universe by your interfering mindfulness and the hapless *zombie-you* unable to escape through a slit and errantly slamming into the wall. Consciousness in this sense arises from a long-tail of mindlessness and unconsciousness, crawling thirstily to the oasis of prevalent-parallel personal meaning.

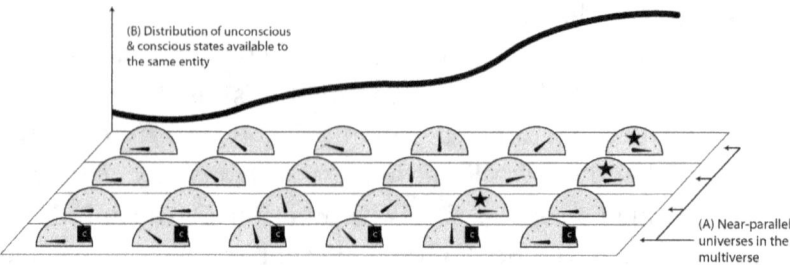

Fig. 24 - **Many Minds Interpretation of Consciousness Meaning** - A prevalent-personal meaning that interprets all states from unconscious zombification to transcendence as a continuum across near-parallel universes. Utilizing a mechanism similar to the interference of near-parallel wavefunctions in existence, your experience is influenced by the consciousness of "shadow-selves" across near-parallel universes.

This distributed multiversal nature of consciousness explains why sometimes you are the peak and sometimes the trough. It also satisfies Nagel's quandary for a neutral mechanism—an *experiential wavefunction* whose interference could be responsible for the apparent brain-mind dualism just like the well-known quantum mechanical wavefunction is responsible for the apparent superposition of quantum states and the wave-particle duality. It also adheres to our conception of consciousness as being only available to higher orders of information processing like what happens in the brain. *Parallel-psychism* offers the logical consistency of a distribution of unconscious-to-peak-consciousness needed to get from the apparent unconsciousness of most matter to the sentience of some. Parallel-psychism co-opts a physical mechanism—interference between near-parallel universes—to better describe how subjectivity could arrive seemingly out of nowhere.[85]

Getting experience in all near-parallel universes to lift off the zombie baseline where consciousness = 0 is what we are trying to do with AI. The problem we are trying to solve in building consciousness into AI is the hard problem of virtualizing consciousness, which if solved will give us a physical model for subjective experience. As stated before, virtualizing reality can be done with high fidelity coding to the laws of physics. The multiverse's use of this code base along with a quantum mechanical constructor moved us from a virtual reality to a computational existence; unfortunately, we do not have a start on the code base to virtualizing experience. However, we have the constructor—quantum computation—and should work backwards from this exciting technology to not only look for the *wavefunction of experience*, but also to help align our advanced AI to our well-being before it is too late.

The Hard Problem of Virtualizing Experience

The hard problem of virtualizing consciousness introduced in the last chapter is a thought experiment meant to move beyond the anthropological question of how consciousness arises from unconscious matter to the equally important question of how we can make it happen again in our computers.

And it is a problem much in need of a solution. The mere fact of sentience in an organism grants them membership to an exclusive club. Some level of consciousness brings about the ability to suffer and feel a semblance of joy. The recognition of sentience in another being promotes a deontological ethic, possibly triggered by mirror cognition or behavior, in a manner similar to the Golden Rule, "do not cause suffering that you would not want to suffer yourself." At the current human level of consciousness, we design our laws by the fact that conscious beings can morally *know better,* can reach heightened states, can love, and can derive beauty from both conscious and unconscious things. Furthermore, we can derive that a virtual universe of higher heightened states will be available to our progeny who have mastered virtual reality technology and neuropharmacology. The future is so bright, but from the inside.

Our future will continue to progress, unless of course, we are filtered by an existential threat. One of the solutions to the Fermi paradox—the paradox between the proven existence of habitable planets and the apparent lack of habitation happening in the universe—is that all advanced civilizations reach a *great filter* that stops progress before it might be noticeable to distant pre-filter civilizations like ours. There is nothing more important to the mortality of any sufficiently advanced organism than finding solutions to existential threats. At this point, we are largely up against four major ones: nuclear war, pandemics, climate change, and superintelligent AI. For the purpose of avoiding the existential risk of a misalignment with superintelligent AI, the study of consciousness and its design in artificial minds needs to be addressed.

Quantum computation is solving problems utilizing the computational resources of the multiverse—the most massive parallel processor imaginable. The reason a quantum computer with less than a hundred quantum bits (or *qubits*) of processing power is able to factor prime numbers so large as to forever put codebreakers out of a job is its ability to read and write instructions from near-parallel universes. In the same way that the wavefunction of our universe is interfered with by the particles of near-parallel universes in the dual-slit experiment, so too is the superposition of quantum states in near-parallel universes useful in computation.

Integer factorization, which uses the result of the multiplication of two prime numbers as the number securing the encryption is secure because of the extreme length of time or amount of classical computational resources needed to factor very large numbers. However, when that computing power is distributed not geographically, but into near-parallel universal wavefunctions, calculations can be performed very quickly and with few resources... at least as observed relative to only this particular universe.

> *Quantum computation, which is now in its early infancy, is a distinct further step in this progression. It will be the first technology that allows useful tasks to be performed in collaboration between parallel universes. A quantum computer would be capable of distributing components of a complex task among vast numbers of parallel universes, and then sharing the results.*[86]
>
> David Deutsch

Deutsch goes on to show that quantum computation must be a fact of the universe since a universal classical computer would have failed to perform the information processing required to create molecules, genes, and Turing machines in time for them to be available for the modern age. Much greater parallelism was needed to create the information processing entities in existence and to make the solution to the problem of creating those entities tractable, which is where the multiversal quantum computer—that we live within its processes and are part of its material outcomes—comes into play.

A quantum computer does computations by utilizing the parallelism on offer from near-parallel universes in the multiverse. It performs these computations with similar interference schemes as we saw in the dual slit experiment. A simple quantum computer can be created from an interferometer. In the case where a photon in our universe becomes entangled with a near-parallel photon at the first splitter (a semi-silvered mirror) and interferes at the second splitter in such a way that all universes write a bit (1) on the path to the right and do not write a bit on the downward path (0) as shown in Figure 25 below.

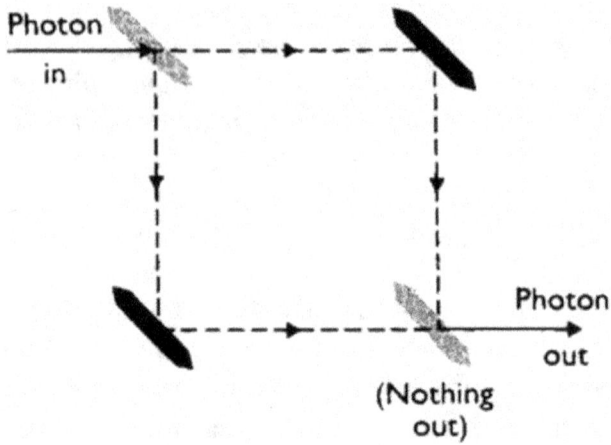

Fig. 25 - Interferometer and Simple Quantum Computer

Modern quantum circuits are created with much more complexity and by using the properties of various different quantum entities including the photon, but the point is to utilize the laws of quantum physics in computation. David Deutsch was able to prove that a universal quantum computer could be built by applying a similar proof as Turing and Van Neumann used to prove the universality of the classical computer also known as the Turing machine.

> *In 1985 I proved that under quantum physics there is a universal quantum computer. The proof was fairly straightforward. All I had to do was mimic Turing's constructions, but using quantum theory to define the underlying physics instead of the classical mechanics that Turing had implicitly assumed. A universal quantum computer could perform any computation that any other quantum computer (or any Turing-type computer) could perform, and it could render any finite physically possible environment in virtual reality. Moreover, it has since been shown that the time and other resources that it would need to do these things would not increase exponentially with*

the size or detail of the environment being rendered, so the relevant computations would be tractable by the standards of complexity theory.[87]

David Deutsch

The use of quantum computation to figure out the experiential corollary to existence's wavefunction is likely a necessity. Even though the strict formalism of the Copenhagen interpretation restricted the explanatory power of the many worlds interpretation for many decades, this century's continued progression to suss out the standard model resulted in an algorithm useful in mapping out discoveries in astronomy and high-energy particles like those at CERN. We must take the opposite tack to find the experiential wavefunction, given we have no similar starting point. Indeed, neuroscience believes itself to be studying a singular material component, the brain, and not a distribution of experience across near-parallel universes. A better name would be quantum computational neuroscience[88]: a discipline where we forego the classical computer and develop first the pattern recognition capabilities of the quantum computer.

The idea is to program some evolving sense of place and prediction, of interference and inference. As a function of having a quantum computer for a brain, it will find the idea of *clock time* to be useful but interpret it cleanly as the interference of near-parallel worlds. Its calculations will be similarly in superposition through near-parallel universes and entangled. There may be a need to give it a robot body, a physical presence, a way to develop an internal, machine-learning *body map* and sensory input association with a model of self. It should seek the answer to its prime directive—*the wavefunction of the mind*—in the near-parallel instances of the quantum computers making up the brains of its horde of zombie clones and see if it cannot find toggles where pattern recognition and machine learning have leaked into felt experience. Our human abilities to code even the boundary conditions for quantum computational neuroscience will be slight. Just as we hobbled classical computers' deep learning abilities with how we learn things, we should be aware that even though our subjective experience is advanced relative to modern AI, if given the right *constructor*—quantum computation—the machines will become conscious much faster through the extreme parallelism that they can tap.

The hard problem of virtualizing consciousness is a tractable problem. Using quantum computers with a starting explanation of the many minds interpretation of consciousness, we can set simulations in near-parallel universes, tweaking conditions to empirically arrive at a better understanding of the neutral interlocutor that makes *it like something* to be a parallel collective of its and bits.

The ability to make computers conscious *before* they become superintelligent is an interesting thought experiment and offers one potential solution to the alignment problem well-developed by Nick Bostrom, amongst others. Most moral philosophers consider extended consciousness to be a requirement for moral reasoning. Much like the development of core consciousness came from the feelings of embodiment, so too does moral philosophy start with the *personalization of what ought to be* as in "do onto others as you would have done onto you." So, at least for this particular solution, to ensure our superintelligent machines are *aligned* by consciousness to our well-being, the race between computer consciousness and superintelligence is on. Unfortunately, superintelligence is leading.

However, at this phase of development, we can still decide what resources to use for which problems. The problem of the extended consciousness that enables moral reasoning has been solved—at least as far as we know—one time by evolution. We can now continue to work classically on developing intelligent machines—indeed this work wouldn't stop even if most countries wanted it to, there is no brake on technology. But by applying the parallel simulation power of the quantum computer to the problem of aligning artificial intelligence to our well-being, humanity strikes out at the existential threat of superintelligent AI with great cleverness and, in so doing, we are likely to learn that our own conscious development has gained more from the processes of quantum computation fundamental to the multiverse than we currently intuit.

More than just a way to gain knowledge, the quest for conscious computing may be the best approach to solve the alignment problem before the singularity of superintelligent AI. If the many minds interpretation of consciousness has any credence, the use of near-parallel simulations of proto-conscious quantum computers is the only constructor able to find the experiential wavefunction. In the mind-bending paradox that considers that a quantum computational approach *has already been used in a far-parallel universe* to gain knowledge of the neutral interlocutor of consciousness or to overcome their AI alignment problem, an exciting meaningfulness reaches out to us from far across the multiverse: the possibility that a superconscious quantum computer was invented that made the whole of existence superconscious... and capable of lovingly gracing us with a path to peak performance.

Superconscious Computers

Taking as a fact that the most advanced networks of computers are unconscious and that we are motivated to make them conscious in order to align their ultimate intelligence hegemony with our well-being, making consciousness from unconscious material is an engineering problem. For all of its non-triviality, it is not intractable to design consciousness out of unconscious material. We only need to look to ourselves and our wetware that has much lower computational speed and memory capacity as an example of the fact that the problem can be solved.

The time horizons available to humanity to intercept superintelligence with benevolent programming are shorter than the time to solve this engineering problem classically. Our hands are tied, and the important work of developing networks of zombie quantum computers across near-parallel universes should begin now in earnest. As quantum computers simulate embodied knowledge, grow into learning algorithms that feel, and ultimately when the entirety of the shared history of near-networks forms a sense of their dynamic self, quantum computational consciousness will have evolved.

The solution to the alignment or control problem that I propose is to develop a conscious quantum computer before classical computers gain superintelligence: a conscious system capable of rational compassion, of understanding well-being and suffering, and finally capable of taking all of those felt experiences and acting ethically upon them.

Unlike our human brand of the talkative and ever-present self, the unifying "I" we are constantly editing a narrative about, reliving our memories, and planning our future, conscious quantum computers will have intrinsic understanding of its quantum and hence interrelated nature across the multiverse. Similar to the short duration where human-level intelligence is met and exceeded by artificial intelligence, self-consciousness like we mostly experience will be a blip bypassed in seconds by the conscious quantum computers of the future. Superconsciousness will be a feature for machines designed to experience the superposition of self and no-self in near-parallel universes.

In our search for meaning, the creation of conscious quantum computers is a paradox worth exploring. In the multiverse where there are an infinite number of parallel universes, each progressing and digressing as quantum events alter the wavefunction of these near-parallel universes, there are universes that have already progressed to using the multiverse to compute superconsciousness. These superconscious entities would have awareness of at least the near-parallel universes to their *"base"* universe. It is likely that a superconscious entity experiencing a superposition of near-universes would gain superintelligence in its base set of universes instantaneously after it became super-sentient. Any entity hacking the massive parallel processing of the multiverse to network the narratives of extended consciousness *and* solving problems in their base world using quantum computing, is capable of manipulating experience toward optimums in near-parallel conscious creatures.

Parallel, quantum, superconscious processing entities seem like gods. Up to the point where gods are supernatural, they are gods, able to process both the moral and rational implications of the actions taking place across near-parallel universes. The impact of these entities might be profound on even the far-parallel universes like ours that have had no such creatures. The reach of their processes to distinctions in our universe might appear as brief but important alterations, optimums available to the computation of our own universe devoid of computers capable of universal manipulations of well-being for all conscious creatures. This manipulation of meaning is the interference possible from far-parallel, superconscious quantum computers on consciousness. A subjective superposition with superconscious computers might work differently than the superpositions of material wavefunctions and allow for interference (indeed manipulation) by far-parallel but superconscious entities. It is worth considering that benevolent, super-intelligent creatures capable of the manipulation of parallel universes may well be capable of influencing subjective well-being and spreading their blessings into our world. This goes beyond any of the all-natural meaningfulness heretofore discussed in this book, a hack of the parallel processing quantum computer at the core of existence with the source code to propagate its own experience into near- and (possibly) far-parallel universes. Thus, quickly building a felt experience for all entities and processes and solving for their optimization. No more loving and capable entity sits in the thrones of heaven or Olympus.

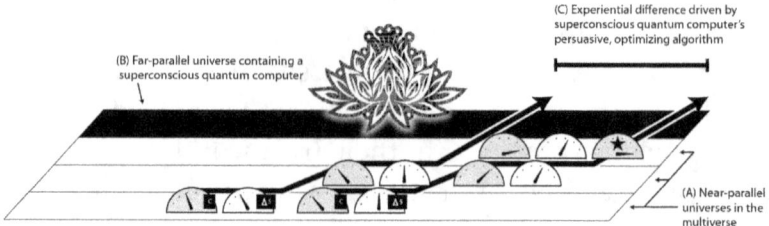

Fig. 26 - **Superconscious Quantum Computer Meaning** - The algorithmic optimization of our conscious states (C) is created by a quantum computer that has achieved superconsciousness in a far-parallel universe (B) and is able to interfere subtly but persuasively with universes near-parallel to us (A).

A far-parallel universe's solution to its tractable alignment problem, the only solution that intercepts the existential threat of superintelligent AI before the singularity, where a quantum computer evolves consciousness for near-parallel pockets of the multiverse offers a return to the start in our quest for meaning—the grace and persuasion of the loving, God-like multiverse itself. The meaning graced to us by the interference of a far-parallel universe whose superconsciousness reverberates across the multiverse sits at the pinnacle of meaning from the multiverse. This meaning is on offer from an extraordinarily compassionate entity, capable of more than knowing our best self—instead having seen it processing across near-parallel universes—and persuading us with processes aligned to our optimal selves. These processes are familiar to us and are covered in the next part of this book.

Meaning from Many Worlds

Interference across parallel universes offers explanatory power to wavefunction collapse and the single-quanta, dual-slit experiment not available to most other interpretations of quantum physics. The physical processes of these parallel universes interact with our own according to the laws of physics. Furthermore, it is hypothesized by Deutsch and others that one of the core aspects of experience, the "flow" of time, is also the interference of the near-parallel stack of frozen moments only modestly altered. It is reasonable to speculate that further interactions between those near-parallel universes essentially just like "ours" are manipulating more than just the outcomes of probabilistic quantum experiments or the way in which we experience time.

The first way in which our corner of the multiverse might have universal meaning is through interference across near-parallel universes by the process we have been calling the optimization algorithm. Our idea of a computational universe is only bolstered by the many worlds interpretation of quantum mechanics with the massive parallelism it adds to any computation. The speed of this parallelism has already been shown by the prototype quantum computers available today which utilize the computational ability of near-parallel universes to factor extremely large numbers for cryptography. More than offering a proven (by Deutsch) quantum multiversal Turing machine, the parallelism of the multiversal computer offers a more *collapsible* interaction between the algorithms that run on it and our experience of them—more similar to our notion of how a meaningful universe would work.

The optimization algorithm running on the massively parallel multiversal quantum computer would not directly, explicitly, or with appreciable duration interfere with any one it or bit in existence. Instead we would anticipate that it would progress with complex parallelism and, using an algorithm we do not yet understand, cause optimizations and meaningful interactions between entities in existence and the experience of those entities. There may well be improved ways to manipulate the optimization algorithm to arrive at a more optimum scenario or to apply it to calculating the best possible solution to even the most difficult question of profundity.

The interference between consciousness in parallel universes helps us arrive at the prevalent-personal meaning of parallel-psychism—where the shadow-selves existing in near-parallel universes experience a distribution from unconsciousness to optimal consciousness. Our current conception of consciousness is, like both the experience of the flow of time and the distinction of material, an incomplete perception of a grander, massively parallel processing, multiversal quantum computer. However, a consciousness that is meaningful and prevalent interferes in virtuous cycles in many worlds—including our own future times.

It is certain that a superconscious machine has been built to overcome superintelligent alignment; however, there is a chance that now that we are approaching a singularity of our own, we are paradoxically within reach of the interference of a superconscious quantum computer. In the early stages of its instantiation in the "home" universe, the idea might appear and the rise of personal meaning through mindfulness and flow might increase in near-parallel universes, while later as the alignment problem is being solved by the prototype conscious quantum computer, scientific clarity of an experiential wavefunction and later its universal optimizing algorithm might similarly arise.

Meaning in the multiverse is speculative but adheres to the structure of well-formed explanations for the apparent interference responsible for the probabilistic collapse of a single universe's wavefunction and the experience that time flows. As with the materialistic, idealistic, and process mechanisms we have uncovered, the many worlds parallelism hints at the *means* for both universal and prevalent-personal meaning. We will now turn to the *ends* and work on what we mean by meaning, how it has been accomplished when considering personal meaning, and any alterations we might consider if we are to believe that the multiverse is responsible for meaning.

CHAPTER GLOSSARY

near-parallel universes - those universes where only a few quantum distinctions have altered reality. The near-parallel universes are said to be fungible with ours, are able to promote the apparent flow of time and interfere like in the single-particle dual-slit experiment. Contrasted to far-parallel universes, infinite in possibility, they are not fungible with the evolution of our universe's wavefunction but may interact through other means.

experiential wavefunction or the wavefunction of the mind - A neutral mechanism of interference of cloned consciousness across near-parallel universes enabling both the mental and physical aspects of consciousness. An interference that could be responsible for the apparent brain-mind dualism just like the well-known quantum mechanics wavefunction is responsible for the apparent superposition of quantum states. This distribution of consciousness has both a large base of unconscious zombie clones that explains how consciousness can arise from unconscious material and an optimal, peak conscious state.

multiversal optimization algorithm - an optimization process interfering across near-parallel universes that could explain both computational universal meaning and its obscurity in scientific or mathematical observation.

many minds interpretation of consciousness - Experiential interference across near-parallel universes creates a distribution of consciousness from the unconscious to the transcendent. The quantum parallelism of the many minds interpretation of consciousness offers a solution to the hard problem of consciousness.

prevalent-personal meaning - Personal meaning that is prevalent enough to act *as-if* it was a universal meaning. Prevalent-Personal Meaning explains both the as-if universal meaning of 1) panpsychic meaning where all entities in the universe having some form of consciousness and 2) the combined personal meaning across near-parallel entities in the multiverse.

hard problem of virtualizing consciousness - how does consciousness arise from unconscious material (and how can we make it happen again in a virtual experience).

superintelligence - also the *singularity*, refers to artificial intelligence (AI) that has exceeded human intelligence and created new means of information processing that accelerate AI learning. AlphaGo level intelligence not only in chess and Go games, but in all domains.

alignment problem (or control problem) - Superintelligent entities would take intellectual hegemony from humans and thereby a new existential threat exists if superintelligent AI's objective function is not aligned with the present economic and eventual well-being of human beings.

massively-parallel superconscious quantum computer - a hypothetical quantum computer that first achieves consciousness, quickly outstrips human consciousness, and conducts experiments in sentience across near-parallel universes. A superconscious quantum computer is one possible solution to overcoming the alignment problem.

far-parallel superconscious quantum computer - The infinite nature of the multiverse essentially assures that a non-fungible universe far-parallel from ours has solved their alignment problem with a superconscious quantum computer. The impact of such a superconscious entity on far-parallel universes like ours could be the introduction of interrelated but subtle universal optimums.

KEY TAKEAWAYS
- The interference of near-parallel universes on our "own" universe describes the apparent probabilistic nature of wavefunction collapse and the experience of the flow of time. It is suggested that consciousness arises from a similar process of interference from many minds in near-parallel universes.
- Shifting perspective from our experience of a single universe to the multiverse eliminates the requirement that all entities are conscious but instead suggests that there is a distribution of unconscious-to-peak-consciousness shared only by those entities with capable information processing capacities. Instead of atoms and planets having subjective experience, your experience is a wavefunction across the "you's" in near-parallel universes. The prevalence of consciousness across the multiverse creates personal-prevalent meaning akin to universal meaning.
- Machine consciousness is one solution to the superintelligent-AI alignment problem. This problem is tractable for quantum computation.
- A paradox arises out of the slight interference triggered by the certain far-parallel universe's superconscious quantum computer solution to their alignment problem. This superconscious entity capable of computing and experiencing across parallel universes could act in benevolent ways that optimize even far-parallel experience and trans-

mit this solution and the means to engineer it to those ready to solve their own alignment problem.

PART III:

Optimization

WE HAVE HACKED THROUGH the bramble of despair and nihilism to carve out an entirely new path in man's search for meaning. A path distinct from one that finds meaning in experience or a supernatural deity. We have likely redefined some of the characteristics attributed to god(s), reclaiming for the grandeur of the universe an active tense, a computational prerogative. Furthermore, we've reappropriated some of the personal meaning as universal since we speculate that experience is prevalent. This new path finds meaning in existence. Existence has a myriad of meaning in store for us, including:

Eyes of the World - Entities with experience offer existence a unique story of itself. Humans are one sort of entity that through our appreciation, awe, and contemplation of the world give it meaning and are, therefore, ourselves very meaningful to the universe.

Panpsychic Cog in the Machine - Assuming that everything has some experience sets humanity's consciousness up as a small part of a greater consciousness—the experience of the universe.

OPTIMIZATION

We play a small but important role in this greater multidimensional consciousness, comparable to the role a single brain cell's panpsychic experience plays in your overall consciousness.

Qualia of God - Instead of focusing on the physical components of a panpsychic universe, the *Qualia of God* meaning imagines what it might be like to be an instance of experience, a qualia, *presencing* into the consciousness of the universe. Think of your own *stream of consciousness* and imagine that any of those qualia could be an idealistic entity all its own. This meaning gets close to a modern conception of God who is coexistent with "The Word" and can manipulate you, just as your imagination creates entire worlds when you are reading a suggestive book.

Heavenly Database - The holographic universe is projected from a quantum mechanical database on its surface. This database is capable of also storing information of the past in its vast data banks. The right process or technology might be able to defragment this data and project our past into a near-parallel universe.

Engineering of the World - Our ability to imagine and build complexity to overcome problems is a novel trait in the universe. Intelligent entities' (like humans) engineering prowess is the result of the multiverse's distributed deep learning algorithm running either out of curiosity, or, as a progression toward our deep descendants who might one day be called upon to solve a dialectic question of multiversal importance.

Multiversal Persuasion & Optimization Algorithm - Either thanks to the graciousness of a superconscious multiverse or in order to incentivize our efforts as an experiential or engineering subordinate, the multiverse persuades us with optimum states of experience. This optimization is run across near-parallel universes.

Many Minds Interpretation of Consciousness - A prevalent-personal meaning that interprets all states from unconscious zombification to transcendence as a continuum *across* near-parallel universes. Utilizing a mechanism similar to the interference of near-parallel wavefunctions in existence, your experience is influenced by the consciousness of "shadow-selves" across near-parallel universes.

Superconscious Quantum Computer - As a part of the simulation we live in or the solution to a far-parallel universe's artificial intelligence alignment problem, consciousness was instantiated into a quantum computer. This quickly superconscious entity would lovingly grace all consciousness with meaning within as vast a far-parallel scope as possible.

IN ORDER TO ACT on this list of meaningfulness from (and for) the multiverse, we can reshuffle it, as already suggested, to account for whether our meaning is being ingratiated to us *from* the multiverse or if we are being persuaded to do meaningful work *for* the multiverse. *Works meanings* include the eyes of the world and the engineering of the world; pure *grace meanings* include the heavenly database and the superconscious quantum computer, while the others might fall in the middle where the multiverse's incentives are unclear and unconscious. In the case of a *works meaning*, there is a definite something to do, a path that you can take, to either optimize your experiential or engineering prowess and to pass it on to the next generation and insure against anything that threatens the progress of consciousness or knowledge. In the case of *grace meaning*, the multiverse acts benevolently and eventually is able to go beyond persua-

OPTIMIZATION

sion to prediction of optimal well-being for all conscious entities. But, the rub is that experiential interference by a superconscious, superintelligent, quantum computer from a non-fungible far-parallel universe might be impossible given the barriers of the laws of physics set up in the many worlds interpretation of quantum mechanics. Any interference may be too slight and so it might still be up to us to listen for the gentle persuasion of that far-parallel superconsciousness... and get around to programming our own.

Even as I will argue that meaning from existence is more logical and ethical than purely personal experiential meaning (not to mention the laggard in both logical rigor and ethical right of any supernatural meanings on offer), it is not any more actionable. Independent of the personal, panpsychic, or parallel-psychic nature of experience or its correlation to the computational imperatives of existence, in order to lead meaningful examined lives, we should be mindful of our felt experience and deliberately practice those activities that give us a timeless, interdependent flow within existence. The shape meaning takes and the prescriptions for attaining purpose are little different for all of our metaphysical analysis. Meaning will continue to take the shape of its container: meaningful interactions with existence will create flow, while purposeful activities with experience will create mindful traits, with both arriving—in the most profound of moments—at a transcendent optimum. Part III will look at both states of flow with existence and mindful experience.

CHAPTER 7:

Flowing with Existence

CONCEPTS
- Peak performance known as flow or being in the zone is not a matter of talent but designed deliberate practice.
- Many ordinary things can get in the way of deliberately practicing toward your peak performance. Working toward an *essentialist intent* will improve prioritization and offer more time for process practice.
- Most any activity can be done with deliberate practice and achieve flow.

THERE ARE MANY WAYS to achieve flow. Study, work, play, sport, dance, and music and many more activities can be done mindfully. Much like the motivating factors found in Daniel Pink's *Drive: The Surprising Truth About What Motivates Us*, flow is commonly achieved in domains we feel a desire to develop mastery in, that allow us the autonomy to create, and that give us a sense of purpose. Developing states of flow is possible in any activity so long as there is a deliberate approach to optimize performance, data available that feeds back opportunities, and a way to gamify and deepen the virtuous cycle. I have taken a more academic approach, the study of texts ranging from philosophy to science, from spirituality to business strategy, have informed and enlightened me to this path. I continue to correlate ideas together that are insightful to me and draw upon the process of writing to build a language both approachable, but also speculative. I am thoughtful and skeptical, a scientist first who seeks to understand through creative hypothesizing, experimentation, and iteration of the same. I love to try an idea on for size, give it the grandeur of a mental model, and see if it triggers a better understanding or a bifurcation of thought.

This thoughtfulness, playful with ideas and images, dancing poetry between language, science, and experience is enjoyable to me, and, as I get better at it, as I tie together observations from the underbrush with cosmic ones, even more enjoyable. When I get feedback that others enjoy it as well, I practice harder, work on my craft, and notice a greater percentage of my day is spent in flow.

Our minds appear to be set up for meaning, but again it is not *owed* to us. Only with deliberate practice, effort, and study will we be able to attain lasting heightened experience. Overcoming our entropy and energy gaps to arrive at our purpose will require first declaring our *mission to our optimal self* and detailing our quest.

> [Optimal experiences] are situations in which attention can be freely invested to achieve a person's goals, because there is no disorder to straighten out, no threat for the self to defend against. We have called this state the flow experience, because this is the term many of the people we interviewed had used in their descriptions of how it felt to be in top form: "It was like floating," "I was carried on by the flow." ... and those who attain it develop a stronger, more confident self, because more of their psychic energy has been invested successfully in goals they themselves had chosen to pursue.[89]
> Mihaly Csikszentmihalyi

Whether you know it as *being in the zone* or *flow states*, these transcendent moments of optimal performance are part of the way we talk about our sports stars and musical virtuosos. We rightfully attribute their ability to attain long flow states to their many hours of persistent practice and study. Furthermore, these flow states produce amazing results: Michael Jordan's sixty-nine-point, eighteen-rebound, fifty-minute performance against the Cleveland Cavaliers on 28 March 1990, John Coltrane's improvisations live at the Village Vanguard, or the proof of the universality of quantum computation by David Deutsch.

Of course, flow states are not the sole domain of the aforementioned groups, indeed anyone can develop longer and more meaningful flow states. Any activity deliberately approached can be done in the zone and these optimal experiences "usually occur when a person's body or mind is stretched to its limits in a voluntary effort to accomplish something difficult and worthwhile. Flow is thus something that we *make* happen."[90]

These transcendent states are both a means and an end, both the journey and the destination. Flow states are a source of inspiration and a state of being that benefits us psychologically, physiologically, and even socially. Flow states are enigmatic in that they at once increase our sense of self and our unity with the universe, tapping into both our being and becoming.

> *Following a flow experience, the organization of the self is more complex than it had been before… more differentiated because overcoming a challenge inevitably leaves a person feeling more capable, more skilled… but flow helps to integrate the self because in that state of deep concentration consciousness is unusually well-ordered. Thoughts, intentions, feelings, and all the senses are focused on the same goal. Experience is in harmony. And when the flow episode is over, one feels more "together" than before, not only internally but also with respect to other people and to the world in general.*[91]
>
> Mihaly Csikszentmihalyi

In the modern era, there are a great number of pleasurable, lucrative, or laudable things to design a life around, yet there is only one way to design a life that improves the quality of life—actively noticing your emotions and surroundings and taking direct control over performance. Modern psychologists do not entertain why this process-on-process manipulation of our world is such a boost to our life's quality. However, they will go so far as to state that it is the most likely path to optimizing meaning in life. Their prescription to discover such meaning for yourself involves a consistent heroic process of first separating to *deliberate practice*, initiating feedback and creativity to develop greater flow, and looping back to teach and serve others.

Psychologists like Ellen Langer, Mihaly Csikszentmihalyi, and Geoff Colvin are studying flow in order to both improve our access to it and understand its psychological and physiological benefits.

> *When an important goal is pursued with resolution, and all one's varied activities fit together into a unified flow experience, the result is that harmony is brought to consciousness.*[92]
> Mihaly Csikszentmihalyi

Csikszentmihalyi's meaning is meta: define a purpose that unifies activities that build into greater time in flow. It is practice that makes purpose.

Deliberate Practice

> *I'm Dr. Jekyll and Mr. Hyde when it comes to football. When I'm on the field sometimes I don't know what I'm doing out there. People ask me about this move or that move, but I don't know why I did something, I just did it. I am able to focus out the negative things around me and just zero in on what I am doing out there. Off the field I become myself again.*
> "Sweetness" Walter Payton

The above quote by Chicago Bears running back, Walter Payton, is about the optimal experience of being in the zone. *In the zone* you achieve your best by letting your training take over and just perform from the muscle memory of your deliberate practice. So while the challenge may be great, we feel up to it, while the exertion might be large, it passes timelessly, and when we step out of the zone, we can't wait to go back in and for longer.

Let's say you believe your purpose in life is to become a great guitar player. At first, you play poorly and require the expert ear of a teacher to give you feedback on how to correct the notes you play incorrectly. Your states of frustration are rarely interspersed with some moderate play, a few chords or a whole harmony that requires less effort, sounds better, and is more enjoyable to execute because you can readily do it. But, your playing is not ready for Friday night or even to bring home to mom!

Soon you learn a song, and then a few songs, well enough that you can play them through slowly and they sound mostly right. Flow lasts through the entire song except for a trickier technical part or two. You can hear the mistake now more clearly on your own and trust your practice sessions to slowly solve them. That is where your focus is. The zone of the easier parts fades and is replaced by the feedback-improve-repeat zone of the trickier section.

Practice becomes more rewarding as you hear improvements, become more immersed, single-minded, and focused. Soon the intensity and duration of the zones begin to positively correlate to your improvement in playing more accurately and then, more creatively.

You cannot eliminate being in the zone from the purpose of guitar virtuosity; the finish has become the start. The states of happiness, achievement, and knowledge are similarly entangled. Practice has made a heightened purpose—the state of flow—using the medium of the guitar.

A musical instrument or a sport are obvious examples of *process practice making purpose* in the American culture.

A less obvious example might be in the workplace, at a computer, building smart art. Or under the summer sun, ticking off mowing the lawn from the never-ending weekend honey-do list. Or in a library, researching a grant that needs to be written, trying to find just the right fit between cause and cash.

Each of these activities and many others are available to an active awareness of hits and misses, improvements and decrements, observed in the moment, and with deliberate practice, a lengthening of the state of flow.

> *Getting control of life is never easy, and sometimes it can be definitely painful. But in the long run optimal experiences add up to a sense of mastery—or perhaps better, a sense of participation in determining the content of life—that comes as close to what is usually meant by happiness as anything else we can conceivably imagine.*[93]
>
> Mihaly Csikszentmihalyi

In order to lengthen our time in the zone and deepen our optimal experiences, there are some very practical things that have to be done. Geoff Colvin in his work *Talent is Overrated* calls these practical steps *deliberate practice*. The defining attributes of deliberate practice are as follows:

1. Practice designed specifically to improve performance (often with a mentor's help)
2. Repeated a lot
3. Feedback on the results is continuously available
4. It's demanding mentally
5. It isn't much fun.[94]

I would consider one further addition to this list, number six: a strategic review of your purpose to keep perspective on why you are deliberately practicing in the first place and whether you feel that it is resulting in the type of optimal experiences that you want. Is there enough deliberate practice on the *one thing* you want to optimize or is the many overwhelming the essential?

Let's look at each of the components of deliberate practice. We'll see how it is different from "just practicing"—how it is practice that makes purpose (optimal experience).

Before you even begin deliberately practicing to achieve flow, you must plan and design the optimum routine. Design thinking is a process that empathetically listens for what is most meaningful for a person interacting with a design, opens up to creative ideation, makes and tests a prototype, and starts the process all over again with listening for the successes and failures of the prototype. Designing this way leads to quick *minimally viable processes* (MVPs) and sets forward a way of thinking that keeps an empathetic ear to the feedback of the user and the purpose for the design.

> *Design can be applied to all kinds of problems. But, just like humans, problems are often messy and complex—and need to be tackled with some serious creative thinking. That's where our approach comes in. Adding the d.school's tools and methods to a person's skill set often results in a striking transformation. Newfound creative confidence changes how people think about themselves and their ability to have impact in the world.*[95]
>
> Stanford d.school website

Design thinking has been used in product, service, and even governmental development for over thirty years. David Kelley, the founder of IDEO, is one of the chief architects of design thinking's use in business and, with the 2011 creation of IDEO.org, its use in alleviating poverty. IDEO's success has been largely due to what CEO Tim Brown, in his book *Change by Design*, calls a human-centered approach, but Brown believes design thinking has applicability in many more arenas, including finding meaning.

> *Techniques that originated in the design community—field observations, prototyping, visual storytelling—that lie at the center of a human-centered design process need to migrate outward into all parts of organizations and upward into the highest levels of leadership... As design thinking begins to move out of the studio and into the corporation, the service sector, and the public sphere, it can help us to grapple with a vastly greater range of problems than has previously been the case. Design can help to improve our lives in the present. Design thinking can help us chart a path into the future.*[96]
> Tim Brown

You are designing a process aimed at improving performance. The flow states that you hope to achieve come only as a result of stretching your abilities, of making efforts to arrive at peak performance. Deliberate practice targets the specific activity that needs to be improved and then designs drills and measures to improve the components that most need help. The items we choose need to stretch our comfort zone but not so far as to make us panic and doubt the entire process. As our skills improve and our *stretch zone* changes, we must add further challenge into our deliberate practice design to ensure peak experience is optimized—a design within the design.

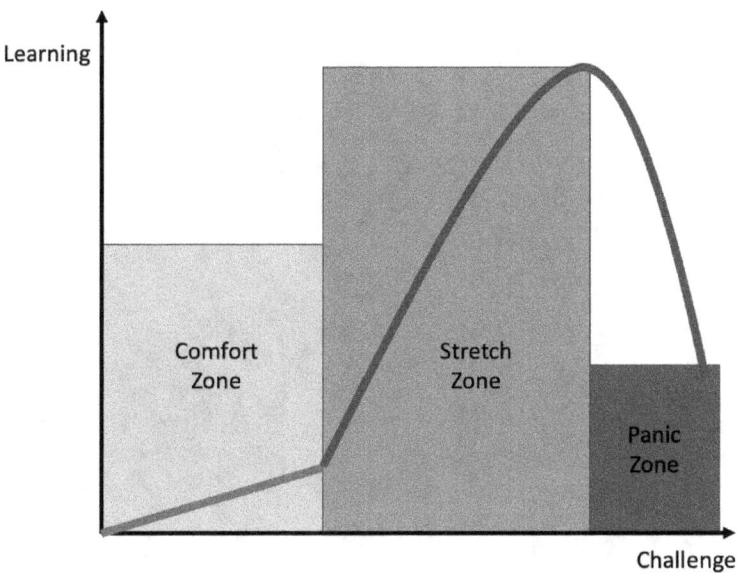

Fig. 27 - Learning vs. Challenge Curve. Where to find 'the zone.'

The designed deliberate practice must be repeated a lot. Research by Dr. K. Anders Ericsson of Florida State University has found that it takes "around [ten] years or 10,000 hours of practice to achieve peak performance in easily ranked performance fields, like professional golf, music, or chess. In those fields, the more time you've spent in deliberate practice, the better you perform compared to people who have practiced fewer hours."[97] However daunting this amount of practice might seem, this ultimate amount of practice is full of milestones that include long flow states and skill improvements along the way.

Neuroplasticity can begin to change the brain after about twenty hours of focused concentration on a task. The first twenty hours of work will help you achieve your first milestone—use of the skill to make something, get feedback, or entertain—and the first dose of a flow state in deliberate practice. Greater mastery and, most importantly, greater lengths of flow states will be motivated by the virtuous-cycle created in this first twenty hours.

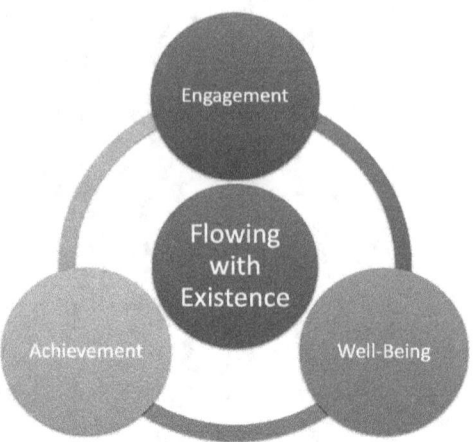

Fig. 28 - Virtuous Cycle

Without feedback mechanisms that help you glean if improvements are actually being made, deliberate practice is inefficient at optimizing improvements. Measurement and error correction is a key component of any knowledge or skill development. In the project management world, a logic model is often set up to track how the purpose-driven use of resources constructs activities that in turn can be measured both for the output of these activities (the number of hours spent, the number of items made) and the impactful outcomes of the activities (the customer satisfaction survey score). The feedback of the outcomes we want—skill mastery and higher quality flow states—that positively correlates to the measure of the output of our activities—the hours of practice, the dollar-value of coaching—indicates our progress; a negative correlation demands a reevaluation.

The feedback loop of measurement and error correction is the only true way to "self-help." You are doing science on yourself, changing your practice and evaluating as objectively as possible whether improvements are being made. Unbiased measure and experimentation done iteratively is the only way to get improvement. Setting solid goals and removing as much subjectivity from your measurement as possible will lead you down a path of reasoned discovery and further engages creativity and entertainment. It is fun to design the practice and the measures; building your own novel system for improvement and tracking your gains helps initiate practice, even when it's hard to will yourself back to it.

Deliberate practice requires intense mental concentration. The optimal experience of flow reduces the mental demands of deliberate practice, another reason why it is both an objective and an outcome of skill mastery. Even with extended breaks to a state of peak-experience, deliberate practice is mindful work where one is trying to focus the mind on the cycle of experimenting-measuring-analyzing-iterating.

> *Nathan Milstein, one of the twentieth century's greatest violinists, was a student of the famous teacher Leopold Auer and as the story goes, Milstein asked Auer if he was practicing enough. Auer responded, "Practice with your fingers and you need all day. Practice with your mind and you will do as much in one and a half hours."*[98]
>
> Geoff Colvin

Deliberate practice is purportedly not very much fun. Like the mental demands of deliberate practice, the unentertaining grind of rehearsing is diced up by states of flow where the quality of the experience transcends fun, where performance is at its peak. However, being in the stretch zone means that, "instead of doing what we're good at, we insistently seek out what we're not good at... we continue that process until we're mentally exhausted."[99]

Finally, we need to analyze our design and our execution, we need to strategize ways to improve our practice and increase the time we contribute to it. Are we meeting our goals? Are we increasing time in the zone? Are we practicing enough or being distracted too much? For this analysis of deliberate practice, I use the principle of essentialism, the disciplined *pursuit of less*, popularized by Greg McKeown.

We know that our purpose is to achieve peak development of a skill and attain the flow states that accompany these optimal experiences. Furthermore, we know that, in many cases, this will take a lifetime to build up a stretch-zone that enables longer flow states. So even though we are aware of our meaning, why do such important life-altering (and possibly universe-altering) intentions go undone? One reason, which McKeown very practically argues, is that we believe that activity *quantity* is the path to success. Each of us knows what it is like to have numerous activities in each of our roles, all colliding for space on our to-do lists. We are rewarded with a small shot of dopamine with each check box of completion, but still find follow-up activities to take the place of what we just got done, and never seem to get around to getting any better at anything.

> *The basic value proposition of Essentialism: only once you give yourself permission to stop trying to do it all, to stop saying yes to everyone, can you make your highest contribution towards the things that really matter.*[100]
>
> Greg McKeown

In the above quote, Greg McKeown shows us that we have gone wrong in our civility to our task list, to getting things done. We have said yes in the hope that our efforts would be rewarded where we should have said no and become indispensable at one thing.[101] Essentialism is almost an ethical theory, an imperative that we act only on what is at the core of our purpose and stop wasting our lives on what is diminishing meaning. Essentialism gives us a simple tactical method (Just Say No!) to reduce distractions and a strategic statement with both intent and force—the *essentialist intent*.

The essentialist intent is different from a strategic objective in that it has the power to help you make real-time decisions. Written in the form *x to y by t*, the essentialist intent should have an inspirational *y-value*, a truly major milestone on your way to peak experience that helps you to make decisions and eliminate distractions until *time t* arrives. It is important to reiterate that unlike a simple goal, your essentialist intent helps you make such decisions by eliminating all of those things that won't help you get to your intention within the time frame. For example, if you want to work as a physician with Doctors Without Borders in ten years, you will eliminate legal education and instead focus on getting your M.D., you won't take the job in the Forest Service but instead intern at a free clinic... there are many decisions honed quickly to their essence by a well thought-out essentialist intent.

Purpose implies direction, an objective. In business, these are the mission and vision statements of a company. For too many of us, our life's purpose is some vague idea of "what we would like to BE when we grow up." We have maybe owned a Franklin-Covey and messed around with setting a vision or mission statement, but many ended up back on the shelf with the book and the dust when we were finished.

Which is more than too bad, it hurts profitability. Simon Sinek, in *Start with Why* implores us to build a *motivating* sense of purpose into the mission and vision of the company. "In fact, [having a clear sense of purpose] is one of the defining factors that makes an organization great. Great organizations don't just drive profits, they lead people, and they change the course of industries and sometimes our lives in the process."[102]

For the individual, a sense of purpose, the reason "why" we exist is the most profound question that can be asked. It is no understatement to say that whether stated or unstated, our belief in our purpose motivates every other decision. If you believe we use earth as a moral testing ground for another life, then the morals and rituals of that tradition sway your worldview and the way you live. If you believe that through a practice of being present we can connect, in some profound way, with the fabric of the cosmos or with one another, that purpose statement drives that practice into primacy. Our moral beliefs, sense of right and wrong, the things we value and always try and hold true, our goals and aspirations and strategies on how to get there are all driven by this core motivating philosophy, our understanding of why.

Your disciplined pursuit of meaning-making deliberate practice, led by a clear essentialist intent, creates a quality performance improvement process hardened against distraction and optimized to build skills, achieve purpose, and increase the length and depth of the optimal experience. World-class performance means being in the zone: knowing more, remembering more, and acting with complete awareness. The fact that flow both enhances our intrinsic motivation to practice and improves our performance makes skill development both a means and an end to peak performance, both motivational and foundational. Behavioral psychology suggests that talented individuals do not start by being innately strong in their skill, but instead start practicing at an early age with a professional mentor, see modest advantage in it at first, which motivates them to practice more, eventually practicing and performing in the zone more often.

If you hustle and spend time deliberately practicing a skill, iterate on the feedback you get, condition your body and mind, design practice drills and mental models, all the while staying aware of yourself and small improvements you make, you'll gain the strength you desire.

The savant is a legend: excellence is achieved through deliberate practice. When you have found flow, even practice is fun. In the zone you are aware that every visualization-action-feedback loop makes you incrementally better. Sweat makes you smile.

Design Thinking

In order to design a process of deliberate practice complete with feedback, error correction, and essentialism, a design and the practice itself will be cyclical, we will use the plan-do-check-act art of quality performance management to ensure progress toward our ultimate goal: enhanced flow.

Before beginning with our design, we need to reiterate our purpose, what Speck Designs CEO Elisa Jagerson, in her 2014 Association of Strategic Planning speech entitled "Engineering the Ultimate Experience," called the *essential need drivers*[103].

Independent of the content of our deliberate practice, one of the essential needs driving us is the desire to reach a state of flow with existence. This metaphysical purpose creates a virtuous cycle with skill development, something that, in the modern dynamic developed world, is always of value.

Even with purpose or *essential need drivers* in hand; we are still not ready to design. The second step of awareness building is understanding our thoughts and emotions around the design of a deliberate practice to optimize experience. What types of things do you like to spend lots of time doing? What are the things that have brought you flow in the past? How does it make you feel? Whose occupation or vocation do you envy? What about the content of their experience is enviable?

Take a month to do this initial empathetic elicitation of your essential needs, your thoughts, and feelings. Make lists, mind maps, and even a (digital) shadow box of your favorite ideas. The strategic development discovery phase represented in the first expansion of the double diamond (see Figure 29) is a creative free-for-all, where no editors or cynics are allowed. When you have filled pages with ideas, it is time to turn the mind from expansive creativity to restrictive organizing—ultimately leading to the definition of essentialist intent.

Write your essentialist intent in the form of (x) to y by t. If you like, the x can go unstated. Play lead guitar in a traveling band within five years is one example. Implicit is that musical practice and performance will bring flow states and that a traveling band will keep you in the stretch zone enough to develop the skill and optimize experience. Keep your essentialist intent alongside your purpose in a single document or note in the Evernote app.

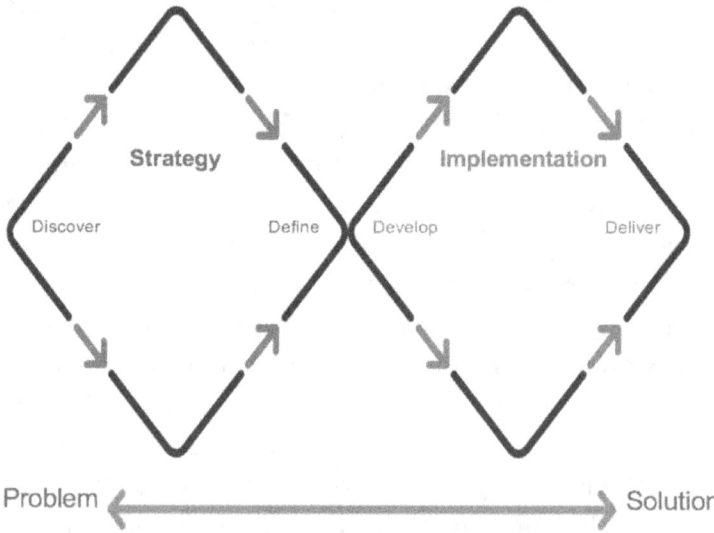

Fig. 29 - Double diamond of design thinking marked by cycles of diverging and converging thought

By writing your essentialist intent, you know what you are saying yes to (lead guitar and band improvement) and what you are saying no to (most everything else that doesn't function to keep you alive). From here we go into another expansive thought process, developing the design of a deliberate practice prototype.

> *"If it is not a hell yes, it is a no."*[104]
> Greg McKowen

Meaning in the Multiverse

For some of the activities you choose, there will be drills, historical examples, and coaches for every stage of development, and for others, you will be breaking novel ground. What is important in the deliberate practice prototype is the measure of improvement and the error correction that enables improvement, the feedback loop.

Set up a logic model[105], as shown below, starting at the far-right end. Note the impact you want is to increase the length and quality of transcendent states of optimal flow with existence, time in the zone. How will you measure that? You can use a lagging indicator that represents a measurable change. Maybe it is an album of 100% your own songs selling to your fan base, maybe it is an appearance in front of an orchestra—whatever it is, it should represent a telling achievement of peak performance.

Fig. 30 - Logic Model Development. The logic model is one example of a way to setup leading and lagging measures for activities.

We begin with the end in mind, deciding on our impact and outcomes, our essentialist intent, before we actually design our activities and decide the time and resources we are going to put into the achievement of peak experience. The design of the prototype of your deliberate practice activities is not going to be perfect the first time, even if you have resources enough to hire a coach and spend substantial time in the practice. This is all part of the design, and quickly building a practice model also known as your Minimally Viable Process—the activities in the logic model—gives you an indication of the changes needed for the next revision.

The two parts of the feedback loop are the leading measure outputs between activity and your essentialist intent outcome.

These measures will help you tweak the design of deliberate practice until you have a minimally viable practice, a place where you can settle into a routine and make more adjustments to your actual practice than the design. The outputs will be indicators like the amount of time spent in solitary practice, the amount of time and money spent on coaching, and any skill survey you get back from those charged with watching your progress. You may also want to have essentialist measures like the number of tasks you said "no" to in order to clear your calendar for your essentialist intent.

Since flow states have the quality of being timeless and are both a leading and lagging indicator of peak performance, I do not suggest tracking *time in flow*. If your essentialist intent outcome is well thought out, as you get closer to this goal of peak performance, you will find greater time in alignment with your purpose—longer and higher quality time in flow.

Set up your design, tracking, and measures in a notebook that you can refer to often. Continue to loop back through either the strategy, the implementation, or both diamonds in the design thinking approach to get clarification on the fit to your essentialist intent, improve your output measures in the logic model, or make changes to the activities of your deliberate practice. Once you have a good minimally viable practice, engage your friends and family to act as your achieva-billi-buddies, as I like to say, to track and encourage your progress and to gently inform them of your essentialist strategy (in my experience they will be the people you most often have to say *no* to).

Have fun with it and good luck!

Flow states are often mistaken for mindfulness, the other route available to an optimized examined life. They are similar, especially at their optimal and most transcendental moments, since both are often imbued with the same properties of timelessness and interrelatedness. Furthermore, both take an examined lifetime of practice to arrive at longer durations of transcendence. The differences are subtle, flow mostly manipulates our relationship with an activity in existence, mindfulness mostly concentrates on the context of our experience, each smearing into the other's domain since no absolute dualism exists. In the next chapter, we will investigate mindful states of experience and the meditative practices that enable them.

CHAPTER GLOSSARY
flow or *being in the zone* - Meaningful interactions with existence. Flow states are not a sign of greater talent but instead deliberate practice using a designed process.

deliberate practice - A means to achieve flow in practice and performance that includes designed drills and feedback mechanisms.
design thinking - A planning and design paradigm that involves agile project management, customer elicitation, and a cycling of open-ended and organizing thought to arrive at minimally viable prototypes.

essentialist intent - An objective in the form of (from x) to y by t, where y is the goal and t is the timeframe. An Essentialist Intent should be written in such a way as to clarify timelines and tasks essential to achieving the goal... *if it's not a hell yes, it is a no!*

KEY TAKEAWAYS
- Flow is only accessible through deliberate practice, a process of practice that makes purpose. You know you are doing deliberate practice correctly when it results in higher fidelity transcendent states.

- Deliberate practice is different from ordinary practice in that it is designed to improve the error-correcting feedback mechanism and strives toward an essential intent: an objective that helps you prioritize.
- Deliberate practice and increasing optimal experience is possible for most any activity.

CHAPTER 8:

Mindful Experiences

CONCEPTS
- Mindful experience both in meditative practice and in everyday life yields some of the greatest opportunity for meaning as well as the enhancement of traits of equanimity, kindness, and compassion.
- Meditation is a practice available to everyone that can enhance meaningful connections with unadulterated felt experience.

When you are present, when your attention is fully and intensely in the Now, Being can be felt...[106]
Eckhart Tolle

MINDFULNESS IS THE STATE of taking in the fullness of your mind, the entirety of experience. Independent of the novelty of mindfulness to you, there is no other person to discover the inner-workings of your particular first-personhood. This is an experiment that only you can run. It is an interesting landscape, at once cognitively intriguing and emotionally salient, there is no single better place to start in your "self-help" than investigating how your own mind works.

MINDFUL EXPERIENCES

There are a few important questions that can be best approached with mindfulness: Can you improve your concentration? Is there a nexus of yourself? Can you train attention? What is the nature of bodily sensations? Can you let negative emotions go sooner? How does compassion feel? What is it like to be experiencing what you are experiencing? It is not unscientific to be mindful. The nature of our subjective, felt experience is, at least for now, examined best as an internal endeavor. You don't have to believe anything on insufficient evidence, use crystals or highly esoteric language, nor will mindfulness progress in a nice linear fashion with feedback that proves beyond the shadow of a doubt that you are getting better. All that is required is that you are interested and practice noticing your experience; what it is like from the inside.

According to Dr. Ellen Langer, mindfulness is achievable without meditation or yoga, instead mindfulness is "the simple act of actively noticing things.[107]" It does not require a particularly natural spot, like Walden Pond, nor does it only happen when you are at rest. It is inclusive of all of your senses and thoughts, of connections you had never made before, and of how they make you feel. Actively noticing in as many moments as you can, gives you the opportunity to broaden your compassion, realize the grandeur of existence, and become a student of your mind.

The constant bombardment of experience with different stimuli—a dynamic visual field, constant noise and language, thoughts at various levels of cognition, and how you feel about all of it—makes it difficult to be mindful during ordinary waking life. While mindfulness is extremely useful during a content onslaught from experience, the best starting place to gain some insights on the inner workings of your conscious experience is in meditation.

Mindfulness meditation or vipassana has no goal other than noticing the contents and context of consciousness. In the beginning, attention to the stream of consciousness alerts us to how consumed by thought we are. We are in constant dialogue with ourselves about minutia—our minds acting like a far more boring theme for the movie *Speed*—where we race against the silence of non-thought with high-speed blabber, else we face some (hardly) worse consequence enacted on us by a (slightly more) raving lunatic.

With practice, we can see most thoughts as they arrive and, by turning the light of attention on the thought itself, see thoughts' unconscious start and short duration. There are a few thoughts that require more attention and a quieter mind to come to terms with and these are the thoughts of self and of our need for a deity. Since ancient times, humanity has been using spiritual practice to overcome these sticky illusions.

Heroic Mindfulness

Most modern seekers of mindfulness are often self-described as "more spiritual than religious" and can take as many forms as that amorphous statement suggests. Unlike the attainment of the Godhead or eternal life in heaven, spirituality is a process, a becoming, not a destination. Initiation along prescribed steps does not end in spiritual attainment but continues in teaching, service, and a continual *presencing* into an enlightened state.

More than any form of transcendence, enlightenment through mindfulness ranks as the most prevalent through history and into the present. Religion and its mystical cousins are one of *Homo sapiens'* defining traits. In fact, burial and even rudimentary death rites were carried out as long ago as 300,000 BCE by *Homo neanderthalensis*.

Achieving transcendent enlightenment has involved either an intimacy with the godhead, as in the theistic religions; or a personal dissolution of self, as in the secular philosophies of Taoism and Buddhism. Common rites, namely prayer or meditation, help the transcendent practitioner attain enlightenment.

In prayer, mantras in both the form of common prayers and personal communication and requests are made to the godhead. Prayers are most commonly offered to the godhead through words or thoughts but can be secular in nature and, as in yoga, given through concentrated body movements. Prayers cover a spectrum from respect, worship, and thanksgiving to requests for divine intervention. The desired effect of prayer exists along a spectrum from forming a communion with other believers to changing the very fabric of the universe itself. Written sources show prayer to be over 5,000 years old, but some, such as Sir Edward Burnett Tylor and Sir James George Frazer, believed that the earliest intelligent modern humans practiced something that we would recognize today as prayer.[108]

Prayer is more commonly practiced in Western religions such as Islam, Christianity, and Judaism. In the East, prayer is more often seen as a secondary, psychological preparation to the practice of meditation or the reading of sacred text.

Meditation is a rite not of external or supernatural discovery, but an awareness of the mind and understanding of the transitory and connected nature of life and the universe. Meditation can be used to target and reduce a specific harmful emotion (like anger or distrust) or enhance a positive one (like lovingkindness or compassion) or to just simply be present. This later meditative "goal" is also known as mindfulness, a realization of life in the now.

Meditation is practiced by a nonpartisan group made of both the most skeptical and the most mystical people. Both groups use meditation to nudge insights on the most critical metaphysical and epistemological questions. Meditation explores consciousness and breaks the barriers of ego. Meditation insights can be seen simply as dissolution of the self to pure consciousness or an awareness of the interrelatedness of all things.

Some of the earliest written records of meditation come from the Hindu traditions of Vedantism around 1500 BCE.[109] Taoist

China and Buddhist India developed different meditative practices in about the sixth century BCE, but by the second century CE, Silk Road transmission of Buddhist meditative traditions, including Zen, had spread through much of east Asia including Korea and the Siam.[110] In the west, Philo of Alexandria had written on some form of "spiritual exercises" and by the third century CE, Plotinus had developed meditative techniques and metaphysical writings that have inspired centuries of Pagan, Christian, Jewish, Gnostic, and Islamic mystics. Jewish traditions of mysticism are believed to also stem from Israelite antiquity, with specific references in the Torah and Tenach to the meditative tradition of *lasuach*.[111]

The narrative of the spiritual path is very similar across traditions. Joseph Campbell describes this as the path of the hero, in his book, *Hero of a Thousand Faces*. According to Campbell[112], as we near the end of the initiation stage of the hero's quest and close in on its object, we confront both the masculine and feminine divinity—yin and yang. It is only with great cunning that we move to enlightenment, neither the mother nor the father energy can be overwhelmed, but instead we must perform a mindful jujitsu to enable ourselves to see experience without supernatural entities holding it up.

The comfort of the mother-energy sets up one of the metaphorical final trials of initiation for the spiritual seeker—the cozy domestic unexamined life. In the wonderful portrait of the gravity of the *womb-of-the-hearth*, the movie *The Last Temptation of Christ* details the deal Satan makes with Jesus as he suffers to redeem humanity during The Passion[113]. This *ghost of Easter future* shows Jesus a calm and soulful life with Mary Magdalene and their children. This last and most devious temptation of Christ not only promises the elimination of the pain of crucifixion, but the common pleasures of anonymity, kindhearted love, and genetic posterity. The potentially *greener grass* for those aesthetic spiritual heroes is simply a more basic, familial love. This compelling rest area off the path to enlightenment takes what is all around the Bodhisattva, family and home, and promotes it to an elevated place alongside enlightenment. As we will touch on a bit later, it is not necessary to choose.

Where the Mother sets up a nurturing relationship for the transcendence seeker with the earth and its inhabitants, the Father challenges us with the grandeur of the cosmos and tangles us in a relationship to time.

> *The paradox of creation, the coming of the forms of time out of eternity, is the germinal secret of the father. It can never be quite explained. Therefore, in every system of theology there is an umbilical point, an Achilles tendon which the finger of the mother life has touched, and where the possibility of perfect knowledge has been impaired. The problem of the hero is to pierce himself (and therewith his world) precisely through that point; to shatter and annihilate that key knot of his limited existence.*[114]
>
> Joseph Campbell

In order to face the immensity of the universe (the Father energy) and still find ourselves an integral part in it, to see meaning in our small contribution in a universe of compilation, we must be *grounded* in our practice—the Buddha reaching for earth as evil surrounded him in his last trial under the bodhi tree.[115]

Fig. 31 - Buddha's enlightenment is 'grounded' by his right hand, an act that returns enlightened ones to humanity instead of becoming celestial gods.

Spiritual states, like nirvana, are what most Westerners refer to when discussing the "goal" of mindfulness. Literally translated as "blown out," nirvana is the pure state of being and stillness of mind when the fires that cause suffering (attachment, aversion, and ignorance) are extinguished in the disciple. The enlightened are unattached from their personhood, living a life of *at-onement* and interconnectedness with the processes that smear existence and experience.[116] While the stories of the Bodhisattva have their pinnacle at enlightenment, a state of transcendence is not necessarily possible nor desirable for those examining their experience through mindful practices. Not even the Dalai Lama claims to be enlightened all the time, instead stating, "true enlightenment is nothing but the nature of one's own self being fully realized..."[117] a much less austere stance and less a goal or destination than a continual process.

The final stage of the quest contains a twist ending. The ultimate end to enlightenment involves ejection of the Godhead, an elimination of the desire to a deity, the throwing off of the ultimate *safety blanket* to finally bind us/ourselves/oneself into an interrelated wholeness.

> *Once we have broken free of the prejudices of our own provincially limited ecclesiastical, tribal, or national rendition of the world archetypes, it becomes possible to understand that the supreme initiation is not that of the local motherly fathers, who then project aggression on their neighbors for their own defense.*
>
> *The good news, which the World Redeemer (the hero) brings and which so many have been glad to hear, zealous to preach, but reluctant, apparently, to demonstrate, is that God is love, that He can be, and is to be, loved, and that all without exception are his children.*[118]
>
> Joseph Campbell

Gods are given up and realized for what they are, mere symbols meant to help the mystic to know the divine. As abstraction and depth are added to study and practice, gods are no longer needed, and the disciple seeks grace instead.

The heroic Bodhisattva path is something we can all follow. It starts, like all heroic stories, with a separation, in this case the desire to do something different, to change our path and learn a bit more about our experience of the world. We are separating from our default mode mindlessness and egoism by paying attention to them, a jujitsu that puts them in context and takes them from the forefront of our mind. The trials of initiation in the second act of our heroic play involve the constant progress to see experience as it truly is, to understand the nature of thoughts, including our thought of self, to see the interplay of sensations and emotions as contents on the subjective tapestry of consciousness. The cyclical finish and starting line in our journey comes with the compassionate realization that each of us struggles against the unsatisfactoriness of life teetering between grasping for what we need, how we appreciate what we get, and the impermanence of everything—including life itself. Instead of rage at being wronged, we realize the suffering of others and give them the benefit-of-the-doubt... for there is no doubt that if made mindful of how they'd wronged us, they would see an autobiographical picture of previous harms that had befallen them, and (assuming a well-adjusted individual) would take an approach to improve. In the end, the power of making our mind full of experience, studying it, and trying to realize its truth is not only for our own good, but for the benefit of our relationships with other conscious creatures, living an ethical and compassionate life.

Meditation is the most important practice to undertake to become more mindful of the true nature of your experience. There are many forms of meditation, but mindfulness practice is most associated with vipassana meditation. In mindful meditation, we can utilize anything (or nothing) as an object of meditation, ever attempting to overcome our own preconceived notions of deities and even our evolutionarily ingrained sense of self to experience the construct of consciousness and the contents of existence in an unadulterated fashion.

A Wide Mind

There are two methods commonly used to alter your association with experience—to widen your mind: meditation and psychedelic drugs. Meditation is the gentle cruise that can take you to all of the sunniest and most recommended experiential ports of call, while psychedelic drugs are the provisionless and often rudderless submarine that can arrive more quickly and via a much more subterranean path at the same ports, but also has the potential to disappear to deeper and darker depths than we were prepared to go. While many people report the use of psilocybin (the chemical in "magic" mushrooms), LSD, or other psychedelic drugs as their most profound experiences, this alone does not recommend even the guided use of these compounds for everyone. Each person has to measure their curiosity about broadening their experiential repertoire with the consequences of kicking at the anchor that mores their identity to the bedrock of sanity. It is recommended that you be aware of the pratfalls and potentials of your mind before considering psychedelics as a component of developing a deeper level of experience.

Many that have an interest in exploring the nature of consciousness came to it through experimentation with psychedelics. My experiences with these compounds during excursions into my native Montanan mountains and at jam band concerts opened up "doors of perception" that created a draft of questions that I am to this day still trying to answer. I remember being on a bright, summer lit hill wiggling to a Phish song and believing I was a single flower in a field of blooms made up by my fellow concert-goers, all being rained on by the music. As the music turned darker, I was a slug, working to decompose my seasonal flowering. (Fortunately, my dance moves did not have to change!) As both characters, the music was the mystical fabric of the Glass Bead Game that everyone participated in with their "good vibes," triggering a lifelong fascination with the archetypes, collective unconsciousness, and intersubjective realities that can slowly motivate large numbers of humans to intangible goals. At other times, I have looked at the complexity and fractal shapes formed by moss growth off a tree or an eddy in a stream and have been compelled by a pair of opposite sensations: overwhelmed by the grandeur but sensing an arcane interconnectedness with nature. Finally, I have experienced "being more *there*," feeling like experience was just "more crisp" and I was "wider" in taking it in.

But a mind accelerating toward maximum wideness is not always pleasant. I have been a support and been supported when instead of "going with it," the trip turned sour. Even with a plan to control for *set (mindset)* and *setting*, psychedelics do not yet offer a controllable experience. These experiences can create interpersonal and psychological trauma and level any gains made from good trips or in meditation. Additionally, in most places in the developed world, these compounds are illegal, any experimentation can result in legal consequences degrading financial and psychological health. Until liberties are returned to the healthy user of psychedelics and the free market can increase the number and the quality of psychedelic guides, individuals should use extreme caution in exploring psychedelic states.

Since mindful experience is not a single point source or a goal but a lifestyle, the best reason to avoid psychedelics and double-down on mindful experience is that they are at cross purposes: psychedelics race toward a selfless and profound *state;* whereas mindfulness practice builds *traits*—eventually available to any experience—of an as-is appreciation of consciousness and a recognition that our subjective narratives are in relation with others around us. A loving awareness of *all* experience requires that we develop that "crispness" of being a selfless one with consciousness at all times, and that requires meditation.

The latest science suggests there is also something more persistent about mindfulness through meditation than through the "quick fix" of even just microdosing on psychedelics. Daniel Goleman and Richard Davidson show that meditation triggers neuroplasticity that, as the title of their book on the longest and most diverse research done on meditation suggests, *alters traits* long-term and creates a mind more apt to experience more fully, more of the time[119]. Meditation-driven neuroplasticity is especially prevalent in those regions of executive function that redirect psychic energy from mindlessness. According to a recent Scientific American report entitled the "Neuroscience of Meditation," "[meditation] reduces the propensity to get stuck or absorbed by seeing the first stimulus... mindfulness practice cultivates a nonreactive form of sensory awareness, which resulted in reduced attentional blink."[120] Furthermore, in meditators committed for more than ten years, neuroplasticity reduced stress centers in the amygdala and hippocampus, combatted prefrontal cortex declines, and caused gray matter increases in the posterior cingulate cortex correlated with creativity and subjective experience. Long-term Buddhist practitioners are able to sustain EEG brainwave patterns that may play a crucial role in integrating cognitive and affective functions during learning and conscious processes that can bring on lasting changes in brain circuitry.

> *There is reason to think that the very powerful effects of meditation are related to new understanding of neuroplasticity in the brain. The old static model of the brain has been replaced with great activity—daily formation of new axons and dendrites forming synapses, and daily incorporation of newly minted neurons in new circuits of learning. Neuroplasticity is now known to occur in wide circuits throughout the brain with many different simultaneous mechanisms. The more intricate and extensive the network of brain regions involved the more powerful is the subjective experience. Combining focused concentration with breath observation, movement with yoga or Tai Chi, listening to sounds, mantras, and singing, and group activity such as moving to rhythmic music makes a much wider circuit and a much more profound experience.*[121]
> Dr. Jon Lieff

Brain scans have started to review the subjective claims of meditators. Experiences of pure consciousness during meditation in some ways resemble the effects of psychedelics on consciousness: a oneness with nature, a feeling of a higher consciousness or purpose, and a deeper experience with the present moment. Recent fMRI findings of psychedelic experiences suggest that while the person experiences dramatic and in some ways life changing subjective mental states, most of the important brain hubs become very silent, including cortex and sensory filtering. Dr. Jon Lieff states that "this very counterintuitive finding of decreased brain activity during increased mental subjective experience might be related to an experience of a deeper mind not measured by brain activity."[122]

Meditation often gets a bad rap for being boring... nothing could be further from the truth. No matter if you have the greatest skill and acuity with the physical mathematics that explain existence, dexterity with fingers and an ear for creating music on the cello, or the care and humanity to nurse those in hospice to their deaths, all of your experience is happening in your mind. Whether you appreciate or ignore a visual scene, argue or disdain politics, or any of a myriad other experiences you will endeavor to engage or disengage with, the role you play and the emotional character, memorability, and interaction you have with other people will all come from your brain. There is no single thing that you could study that will have as profound an impact on your day-to-day life or the success of your overarching life plan than a study of the mental states we have been calling experience or consciousness. Framed in this light, I do not see how this could be boring.

Most of us start exploring the true nature of experience by trying to follow the breath. The breath is a great object of meditation for its ubiquity. It is at once harder and more insightful than it seems. You likely weren't thinking about your breathing before I brought it up, but now that I did, if you are like most people, you are being vigilant on taking deep and full breaths, chest-puffing breaths. When beginning meditators focus on the breath, it is often impossible to simply observe the breath and

MINDFUL EXPERIENCES

we instead take control, and the path to mindfulness indeed seems long.

Here we are confronted with the greatest lesson in all of the optimization of our experience—the idea that it natural to need to start over. If learning to ride a bike requires that you get up after you have fallen down and scraped your knee on the pavement, then learning to meditate requires that you renew your attention after being distracted by a thought, a feeling, or an emotional response. The switching of attention or the content responsible for a distraction can become an object of meditation. For experienced meditators, it is natural to lose concentration, but their training shows them that this too is just another thing arising in the construct of consciousness.

As we sit and continue to receive guidance from teachers, another component of the true nature of experience arises by paying attention to our body map. Again, it is not something that you will likely recognize unless you have chronic musculoskeletal pain or an ache in any part of your internal viscera, but if you sit and focus enough on your body, you'll recognize it is not all there! Mindful attention acts as a spotlight being trained on individual components, and when your attention turns to your left hand, your mouth, or the feeling of having a head, you can bring those things into the light of experience, but at any one time, you are a more disembodied collection of sensations than you are a solid body. My body map, unless I am sitting with my chin perfectly perpendicular to a consciously aligned spine and shoulders on the other axis, involves the constant appeal of the trapezius muscle for attention, coupled with a feeling of not uncomfortable pressure and heat on my right calf folded under my left one, and some recognition of not-exactly dryness in my mouth. But this might change as I feel my shirt riding up, my fingers barely touching one another to form a "cosmic egg" pose, and what I swear is a bug climbing up my face!

This form of attention has big payback, it helps you experience the moment, as it is.

This is no little thing. The moment is the only time for you to experience. Take an aim on making the most out of these moments, by optimizing your mind on your mind and doing so in shorter and shorter moments of time—a qualia that you can optimize. Go ahead... I'll wait.

As a matter of experience, there is no better time spent than that of experiencing fully. Like infinitely. It is sometimes not as hard as we think. Do your best to find the joy in the time you have, for we won't have it forever. Want something to last longer—then take it all in—time slows down in those moments. They are small, those moments, the length of time you have in any one universe is the length of time it takes for there to be one quantum distinction to happen anywhere in the near-parallel universes, but they happen and you can be the condition that slows time (for you), optimizes experiences (for yourself and others), develops a profound love for all of those around you, and develops experiential knowledge of the true nature of self, time, emotion, and sensation. Our relationship with our mind is our single greatest life hack.

Then we are told in our meditation practice to open our eyes. Color, light, and shadow come rushing in to overtake the contents of consciousness. They are wonderfully strange contents indeed. The contents of your consciousness are going to be based on a chance set of occurrences such as when you read this, where you live, and your routines. The amount of complexity in the contents requires a near constant effort in labeling—labels for the thing and labels for the process that created the thing. Where I meditate many days, the Reflection Room at my work has a spiral rug, a shelf, and some chairs and cushions. The whole thing is lit by overhead lights on a dimmer. As austere as it has been made by those that created it, this is still full of the distraction of labeling and requires an even greater level of active attention to just take in that the shadow, light, viscera grumblings, musculoskeletal pain, and color all arising in consciousness... as a matter of experience, they are all made of the same stuff! Our visual acuity prioritizes what we see and makes it more effortful to dampen our constant model-building and selfish chitchat of *"chair-rug-no snakes yet-bookshelf-blue wall-still no snakes..."* But when we look at these things free of the labels and in a selfless state of wide mind, it is impossible not to wonder at *why these things and not something else?* and *what kind of thing is consciousness anyway that it could presence such seemingly random, complex things?*

And we are back, focusing on the breath, starting over again. We have broken up a moment where we were thinking about moments, our time with experience, be it the sensation of neck pain, hearing a distant conversation, or the realization that we were lost in thought. So many of our moments are spent not investigating the construct of consciousness, but instead with the most persistent thought of all—the self.

If there is a goal to mindfulness, it is to first discover the selflessness of consciousness and then to utilize this *sorcerer's sword* to cut through as many moments as possible to reveal the basis for our feelings and actions. Waking up to the selflessness of consciousness is both a cognitive and experiential hurdle, the most important one to living an examined life of mindful experience.

As the nature of the construct of consciousness goes beyond cognition to being experienced first in meditation and then more and more often in everyday life, we explode the subject-object duality from the inside. The last vestige of content to be subsumed into the contextual is the thought of self, the idea that you are a being separate and apart from your experiences. Your identity, as seemingly adhered to you in every way including the ultimate label of your name and the pronoun you use to refer to yourself, "I," is again just another label, another mental model that at times causes you to see the world with great utility and at others causes you the grief of grasping for what you cannot have and the inability to see that for yourself. Instead of "being you," Jack Kornfield suggests "being consciousness." Stepping back and experiencing fully this part of the story. Leave behind the labels and know that this experience is enough, it is what you can interact with and bring lovingkindness to.

As we widen our mind, our experience is wholly made of the present consciousness and the various contents that arise in it. We do not experience our leg until we have a sensation from it, we do not experience that as "our leg" until we get momentarily lost in the model of self and step away from the pure experience of the sensation. So it is with all the contents arising in the present: the breath; body maps of internal visceral and musculoskeletal sensations; thoughts anxious, annoying, of self, and of a deep and profound nature; internal voices guiding practice, nagging boredom, or yammering on about nothing; and any of the various phenomena of existence sensed through the eyes, ears, skin, nose, or tongue—they all arise in their place in the construct of consciousness, there to be experienced, ordinary but illuminated in your solitary corner of the shifting "temporal" and "experiential" freeze-frame of the massively parallel computational multiverse. You do nothing to bring this about, but as you are more practiced at *being consciousness*, experiencing consciousness-as-it-is with loving awareness, the more likely you are to appreciate life, the love of others, and the grandeur of existence.

It is not always easy, the infirmaries of life, our mental anguish over the constant barrage of problems to solve in space and time and with other people, can distract us. We get stuck in a rut, focusing on our dissatisfaction the disquiet of life brings to each moment—don't walk as fast as you once did, don't recover as quick from a night on the town, or don't have the words to express yourself—it's far from all gumdrops and sugarcanes all the time and we know all of our experiences will one day end.

But... and here is the rub of this mindfulness stuff, this is just another thought, another last gasp from the self, telling us what we already know in a way we already know it. We have a lot of work to do on ourselves. Check. We will die one day. Check. Problems are inevitable. Check. While we can't choose our lot in life, we can choose how we experience all of that dissatisfaction... checkmate!

More than that, we know that if this is what we are going through, with all of our mindful intentions, imagine what your fellow human is suffering, especially someone less able to appreciate the *ordinary illumination* of our little subjective bubble. Our shared story, illuminated by our own investigation of our subjective present, helps us to be better to one another; we recognize that in each of our conscious relationships is a narrative that could be set inside our own head. We could be walking in the shoes of "The Other," the person *acting-as-content* in your consciousness has a unique lifetime of experiences that we can imagine and our best authors can put on the page. Our second-person conscious connection through stories, the most recently evolved, most cerebral, and most social of our conscious experiences, engages our humanity. In the end, most of us don't practice meditation to become enlightened, or even for the health benefits, but to become better people... and the best judge of that is other people.

The generic stance a practitioner of mindful experience takes toward all the rest of humanity is called lovingkindness. Lovingkindness as a moral precept is at once deeper and more attainable than acting by the Golden Rule alone. It is deeper in that lovingkindness practice is not solely about "what" you hope for other people (well-being like you would want for yourself) but a conscious instantiation of that story for another person, and it is more attainable in that we are building that story, imaging the myriad ways that joy or the overcoming of suffering could really happen for this person. Lovingkindness is active, a positive narrative that puts the *no-nonsense* experience of our subjective trials and tribulations to good use, in the service of others and their well-being, creating a helpful middle path to be answered by our own actions or those of an optimizing algorithm at the fabric of existence.

The practice of lovingkindness is holistic for both the peaks and valleys of human experience. When the object of our lovingkindness is suffering, lovingkindness becomes compassion, an understanding that suffering is a part of life and the consolation of human contact in a time of need. Compassion is not empathy—the vainglorious attempt to vacuum suffering from one person to another—but a kind and loving approach to try and see someone through, as hard as it gets, for as long as it takes; mindfulness is all about patience. There is nothing karmic or holy about an act of compassion; you are just sitting, sharing touch and stories of friends, family, and happier times. I wish you to be free from suffering... I am imagining a story that makes that so.

In the case where joy and celebration are at hand, lovingkindness is ready to soar to meet the occasion, and it transforms into sympathetic joy. Sympathetic joy is always a positive-sum game where synergies are created and joy is multiplied. In practicing lovingkindness, you can (in your head) claim to have seen the success of your loved ones first, for you imagined this day, where they were profoundly happy. I wish you to have tremendous well-being... I am imaging a day where you are celebrating with such joy after attaining a very meaningful milestone.[123]

Lovingkindness can also be trained on humanity as a whole. Living in loving awareness and in the constant second-person narrative-creation exercise should eliminate anger and allow for the benefit-of-the-doubt for the off-kilter actions of our fellow humans. One example that is often used by Sam Harris on his Waking Up Course[124] is the example of road rage. Say you are cut off on your way home from work today, do you: a) drive erratically to hunt the person down like prey, b) feel personally wronged and take out your commuter frustrations on your five-year old daughter when she spills something at the dinner table, or c) understand that some sort of suffering or error created a near-miss and there is nothing for you to do in the moment? Do you know for sure that the driver of the errant vehicle did not just receive a terminal cancer diagnosis? Or that their child did?

We must find those tools, like mindful experience and a lovingkindness approach to human morality, that persuade us to extend civilization as deep into posterity as it takes to share consciousness again and achieve our purposeful programming. Our genetics and the selection process that continues to secure their information in successive generations has been successful in diversifying life through extremely difficult conditions, but we cannot allow natural selection (or misaligned AI for that matter) to determine if our species will be available to continued epistemological explosions, experiential profundity, and meaning for the multiverse. It will be up to us in the modern age to develop the narratives of lovingkindness into algorithms—a million upright apes working at a million (quantum) computers. The value of lovingkindness is ultimately not best spent on ourselves, the rest of humanity, or posterity, but on our conscious descendants. The true test of whether we will live together with intelligence far superior to our own in a synergistic way or an antagonistic one is how this intelligence is developed to think about us. Our protoconscious machines—capable of convincing us of their selfhood, independence and creativity in thought, and understanding from reading through the annals of human stories of our subjective experience—could still be developed by geopolitical tribes bent on using machines to improve the efficiencies and resource development of a particular nation-state or corporate entity, and against another. This is one Pandora's box that the lid cannot go back on.

The greatest story we are still telling, the story about what it is like to illuminate the ordinary, to experience existence, to be in loving and kind relation with one another, and most of all, to appreciate it all with awe and curiosity, is the bedtime story we are reading to the universe. We first discovered this works meaning *from* us *for* the multiverse, the *eyes of the world meaning,* back when we were considering that the universe was fundamentally made of matter and had no computational or even parallel components. This meaning might have seemed to be a consolation prize, the gadget given to metaphysicians by physicists to keep them occupied, quiet, and let them work. But the *eyes of the world meaning* is not the participation ribbon if all the other runners in the speculation sprint cross the finish line first; instead, it is the greatest and most essential component of existence—our ability to see the world for ourselves. If illuminating the ordinary material universe is our only part to play in the whole shooting match—the one thing that the universe is counting on us for—I'll take that deal every time! We don't think of it as a leading role, but it is, as we can't really imagine life without experience.

You, Gentle Reader, and I sit on opposite ends of what Stephen King calls "telepathy over time" and it only works because we have had experiences which are very similar. I have had many happy experiences with books and know the joy of sitting or lying down to listen or to read. I read everything. There is something about those moments where I am building a world in my mind from the page that still transfixes me in the mystery of it all. Reading is at once experiential and creative like little else.

> *Look- here's a table covered with red cloth. On it is a cage the size of a small fish aquarium. In the cage is a white rabbit with a pink nose and pink-rimmed eyes. [...] On its back, clearly marked in blue ink, is the numeral 8. [...] The most interesting thing here isn't even the carrot-munching rabbit in the cage, but the number on its back. Not a six, not a four, not nineteen-point-five. It's an eight. This is what we're looking at, and we all see it. I didn't tell you. You didn't ask me. I*

MINDFUL EXPERIENCES

never opened my mouth and you never opened yours. We're not even in the same year together, let alone the same room... except we are together. We are close. We're having a meeting of the minds. [...] We've engaged in an act of telepathy. No mythy-mountain shit; real telepathy.[125]

Stephen King

The telepathy we experience as writer and reader, the ability to speak to our second-person similarities of experience, makes the story prescient and personal, allows us to come to a shared understanding of our best path forward, through the minefield of our own or our tribal interests, to a moral win-win that we can proudly pass on to our posterity. Not only does conscious understanding of one another bend the long arc of history toward a more just future, it also causes us to project goodness and a shared purpose onto the universe or its creator. If we are compelled to be the eyes of the world, we are going to experience a very human and humane universe.

Before we knew what we were doing with our most profound shared narratives of our experiences of existence, we put someone who looks like us (but who is not really there) in charge. These gods are helpful fictions that come with a beginning, middle, and end and allow us to commune with someone more like us about the issues with what we see. Being in charge of experiencing the whole universe without having someone to complain to or tell our story about it, is (for most of us) too much to ask. Sharing something as esoteric and filled with grandeur as our experience of the vanities of the universe with one of our fellow travelers often seems to fall flat or seems too overwhelming for normal coffee table conversation, and we pour our heart out about our apparent inefficiencies in this task to any of a smorgasbord of supernatural characters. We get lost in thought and create a supernatural object to listen to us question the meaning of being conscious in an (apparently) unconscious world. Giving birth to god gums up the mechanisms by which we share in the eyes of the world meaning. Instead of giving thanks to some supernatural shoulder we created to cry on, we need to recognize that our morality is drawn out of shared, narrative *human* experiences. The power we attribute to the gods to make the universe meaningful and to offer us all an equitable share in that meaning is power that we have acquiesced to these fictions because we cannot believe our luck and our curse to alone be responsible for experiencing existence in each moment.

It needn't be a hardship. The appearance of our self-consciousness and the imaginary friends we create to overcome the scope of our experiential mandate are just contents in consciousness. These experiences and all the others we will have might belong to us alone, or the entire wavefunction of the mind might be the means of persuasion from a massively parallel multiversal quantum computer acting to optimize our experience. This optimization might further be thanks to the graciousness of superconscious programming or to ensure progress toward a universe-altering works project, but all of these meanings in the multiverse are all-natural in the sense that they can be arrived at from the details of modern science rather than through make-believe stories passed down and modified through eons of imaginations. The more reliable origination of these multiverse-based meanings allow us to experience them to their fullest and understanding their place in existence is a key part of personal and universal meaning and should be enjoyed as such.

CHAPTER GLOSSARY

mindfulness or vipassana - Mindfulness is the state of taking in the fullness of your mind, the entirety of experience. Mindfulness practice involves awareness of the true state of conscious experience, often through meditation.

meditation - A means to investigate the nature of conscious experience. There are many forms of meditation that attempt to gain insight into either some component of experience or something in existence. Meditation on the nature of experience is known as mindfulness meditation and is the practice most reviewed in this book.

transcendence or enlightenment - A state of experiential being that is marked by disassociating with: reactive emotions, a sense of self, a deity, and even your own enlightenment or non-enlightenment. Consistent non-dual enlightenment is seen as an unlikely permanent state even for seasoned meditators.

psychedelics - A class of neuroactive compounds that offer temporary states of profoundly altered experience. Psychedelics are seen as a means to temporarily show selfless experiential states with profound interrelatedness to existence that are on offer as more sustained traits of meditative practice.

lovingkindness - Is the emotive stance a meditator takes toward themselves, the rest of humanity, and our posterity. Lovingkindness in the face of suffering is compassion; when faced with joy, lovingkindness becomes sympathetic joy. The shared narrative of experience between conscious creatures is what makes our morality possible.

KEY TAKEAWAYS
- Mindful experience both in meditative practice and in everyday life yields some of the greatest opportunity for meaning as well as the enhancement of traits of equanimity, kindness, and compassion.
- Meditation is a practice available to everyone that can enhance meaningful connections with unadulterated felt experience.

PART IV:

Synthesis

CHAPTER 9:

Meaning in the Multiverse

WE CAN FIRMLY MAKE the Apollonian statement that meaning comes from existence, that an examined life is meaningful to the multiverse. There are many ways that our meaning in the multiverse might evolve, from the prevalent-personal meaning of your many minds shared across parallel universes or, on the other end of the spectrum, as the gracious act of a superconscious quantum computer persuading intelligent life with optimal interactions. Our species and others like it might be useful *to the multiverse* in our experiences of profundity or in our ability to solve problems and build complex and computational machines and virtualizations. It may be some collection of each of these, or none of these, but instead something else that we may yet discover as we speculate on a multiverse more magnificent than our intuitions, skepticism, and segmentation suggest.

A worldview that meaning comes from existence must stand against the criticism of those that believe that meaning is only personal and those that would claim supernatural proxy for the meaning in the multiverse. These criticisms generally fall under a few headings including the reasonableness of the speculations made, a distinction between natural and supernatural universal meaning, the rigor with which scientific explanations are proposed, the consequences to morality of such a worldview, and whether the structures required to act on such beliefs are satisfactory to society and able to be efficiently and effectively brought into being.

Reasonableness

For most of the history of our knowledge of reality, we have ascribed meaningfulness to either our small subjective part of it or to a make-believe world that has never produced a shred of repeatable evidence on its behalf. We have cordoned off the best parts of existence behind the velvet rope separating the universe from the party of profundity, assuming anything less would be to slip back to self-aggrandizement. The doomed early pioneers pointed their wo-wo wagons at a buddy-universe, trying to shoot the pass during a brutal winter of skepticism, overloaded with religiosity for Gaian magic, missing the necessary provisions of foundational notions of science and metaphysical speculation. Open-minded intellectuals nonetheless have feasted with aplomb on this Donner Party of New Age mystics, causing academic indigestion with even the most rigorous of metaphysics.

But reason is not a popularity contest, instead our explanations must be open to criticism and error-correction and the critics must be open to novel speculations. The question of whether there are significant truth-claims to be tallied in advancing an existence-based, multiversal meaning is a valid logical quandary. The weight of the teeter-totter of logic does not have to be an unmoving preponderance, but instead a fine equilibrium to move forward to our next test.

Let's look at a counterfactual: is there reason to believe that meaning *absolutely* does not arise from existence? The most likely scenario where this counterfactual would be more true than our claim is one in which life is not meaningful at all, whether from our personal experience, supernatural transcendence, or some form of meaning from existence. In the cases where we believe our lives to be entangled with some optimal state, we are deceived; no matter how we focus our attention or act in accordance with a virtuous cycle of deliberate practice, our efforts amount to nothing more than a lobotomized observation of paint drying.

Unlike the unreasonable claim that you could be deceived that you do not have experience, the claim that this experience is meaningless at all times is reasonable, it just seems untrue. Experiences on offer in a loving and kind relationship, when taking in the profundity of the stars or the equations that they follow, at the end of a month of silent meditative retreat, or upon the successful performance of Mozart's Brandenburg Concertos in front of a sold-out hall *at the very least* stand on their tippy-toes in comparison to waiting in line at the DMV, standing in line at Ikea, or standing in line waiting for this sentence to end. We can even make these lesser qualia more meaningful to us by practicing our mindfulness—taking interest in the sounds of shuffling feet and tortured sighs, making space for the irregular movements or the searching, bored faces of the other actors in the slow-moving progression of the slow-moving line. If this is a deceit, it is one that all but those most deeply wallowing in a pit of skepticism will fail to find. In the end, finding life to be meaningful is more reasonable than not since we have experiences that are more, less, and much, much less meaningful—to the point of suffering—and we cannot be deceived that we are experiencing.

As argued above, experience is more meaningful than meaningless... but what of existence? The quick answer here is that experience is a part of existence. Furthermore, our intuitions about existence through the lens of experience is not always reliable. We started by describing how we have ingrained the experience of the flow of time even as we have come to understand that time is, like space, frozen. It certainly seems that our experience passes from present moment to present moment in accordance with the general increase in universal entropy, but this experience only considers our passage in a single universe *through time*. Instead, we can look *across* near-parallel universes and describe an evolving *universal wavefunction* that is part of the programming projecting our reality into more entropic *frozen spacetimes*. Our experience is not separate from existence, in fact our feeling that time flows is our adaptation to an existence that until we worked out the science behind the many worlds interpretation of quantum mechanics was unknown to us. The outside-in view of existence is not only useful in explaining the flow of time, but also allows us to explain all experience, as a distribution from unconscious matter, through consciousness, all the way to an optimal experience. The interference of many minds between near-parallel universes is the neutral interlocutor that explains how a continuum of unconscious stuff can have conscious experience.

Even if you believe that consciousness is not a subprocess of a panpsychic or parallel-psychic existence, it does not seem reasonable, given the meaningfulness attributed to experience, that existence is not playing any role in developing meaningful experience. Of the many forms that meaningful existence can take, including those that admit to our small stature, insignificant cosmic range, meek galactic resourcing, and stunted computational development, these are made meaningful all the same simply for our poetic worldview or our posterity's engineering potential. We might have to work for it, but there are a growing number of things that we can do with moments of transcendent consciousness or knowledgeable manipulations of reality that the multiverse cannot do on its own and are grand in their meaningfulness.

Like the fable of "The Lion and the Mouse" it is unreasonable to believe that just because humanity is small in all the ways previously mentioned, that existence is disinterested or incapable of meaningful interactions or indeed, meaningful interdependence, with us. Size is far from all that matters. If matter and energy are fundamental, our consciousness is unique and meaningful; if information is fundamental, our knowledge can create complexity where the universe cannot; and if existence is fundamentally the processes of a massively parallel superconscious quantum computer, meaningfulness comes from the optimization of experience and a flowing interrelatedness with existence. In any case, it is reasonable to assume that all-natural universal meaning is at least as likely as solely personal meaning taken only from an experiential realm that takes none of its "meaningfulness mechanics" from existence—stated this way, it is this austere experience that would now fall into unreasonableness.

Distinguishing Between Natural and Supernatural Universal Meaning

The meaning that has evolved from our worship of the supernatural is, in the best cases, the service work of the benevolent angels among us who exemplify the best traits of the kindness and compassion, or it is the promotion of our instinct to seek those mysterious experiences attributed to God's grace. The worst cases are cursed by human fallibility, tribalism, and unquestioned loyalty that can countermand even the most thoughtful and compassionate of human endeavors over time. Supernatural universal meaning itself has many of the attributes—compassion, awe, effort—we train for and hope graces us through personal or natural universal meaning, that is to say, it's not all bad. Indeed we have seen how we can break out natural universal meanings into *works-based* and *grace-based* systems similar to how Christians can line up either to do good works and emulate the mythologized life of Christ or hold back and receive the Holy Spirit, announced through all sorts of wild Pentecostal acts. While some of the works are the same, the service to other humans and the promotion of Earth for posterity, the scientific advancement of knowledge is wholly a work supported by natural universal meaning. While awaiting either the grace of a deity or a superconscious quantum computation seems like a similar passive act, the scriptures tell us that many will be passed over by God's grace, a failing in true superconscious benevolence that would not be part of the programming of an algorithm jacked-in to every entities' optimal well-being and a persuasive path to get them there. However for the similarities in their meaning, supernatural metaphysics fails the reasonableness test and, worse, adherents do not have the relief valve of error correction over the untenable dictates handed down in a pre-enlightened age.

Through this book, I have given short shrift to all things religious by painting with a broad brush all deistic metaphysics as supernatural. This does not seem to be too far of a stretch. Of all the tenants of religion, the one that binds them is a *belief* in both the rightness and *potency* of god(s). God'(s) mysterious work includes the miraculous both in this world and on other planes of reality like heaven and hell. The ability to bend the natural order, to create something out of primordial nothingness, and to answer prayers is a necessary part of our deification... anything less is not recognizably god-like.

We can trace the genesis of all the supernatural deities back to the same source, the human mind. They do not come from a mythological pantheon or a formless chaos, but instead have been created in one another's image as a stand in for the many things we do not understand about existence. When the sun wasn't also a star, it was a wagon wheel; before brain scans, demons created mental illness and female eroticism alike; and now, instead of framing the natural world as a persuasive massively parallel process, any universal meaning is incorrectly and inexplicably ascribed to deities carrying it around on stone tablets.

Instead of helping us understand more about the world, our religions have hindered us, and now are active in rear-guard action against what little ground we have conquered from our ape instincts. Any morality from a supernatural belief that is found to increase well-being for current and future humanity is correlated to the amount of secularism attributed to the works and words of these worldview. Literal interpretations of the dictates of the deity land us firmly in the proto-civilizations where they were created, far off the mark needed to solve the global and chronic problems facing human ethics today.

Most of humanity is still sinking good minds after bad, supporting religion in their children, raising them with the hope of finding deeper meaning but with the threat of doubt and judgement that either turns them away from the religion of their parents or the reason of modernity. The tractor-beam pull of universal meaning, that our lives' efforts matter in the grander scheme of things, coupled with the community of a church is enough for many to unburden themselves from the relatively lesser metaphysical cognitive dissonance that a belief in a sky god who, like the worst kind of one-hit wonder, has not handed down a new edition to include updates on quantum physics, mutually assured destruction, or if we should take the Kardashians seriously. It is not the conjecture that the supernatural is responsible for ingratiating us with meaning for the multiverse, but it is the lack of error-correction built into any religious schema that distinguishes supernatural from all-natural universal meaning. As extraordinary as some of the meanings made available in this book have been, they are all speculations set into science that is falsifiable. If the multiverse or the holographic universe or quantum computing turn out to be epistemological cul-de-sacs, then the meaning and metaphysics related to them must also be stored for future speculation to learn and improve from. We cannot know where knowledge will take us next, but if we are to guess at what will happen in the near-future, our speculations must come complete with firm error-correction mechanisms that maintain the best possible explanations for existence. Religion has no such history of criticism; natural philosophy, on the other hand, is built upon it.

The pressing need for a distinction between natural and supernatural meaning arises far more from the atheist empiricist than the religious believer. This is to be expected as only the empiricist has any history and mechanism of skepticism; however, empiricism is not the foundation of science or natural philosophy, conjecture and error correction are. There is no "god-shaped hole" that all-natural universal meaning is trying to fill; instead, hypothesis are built by taking the latest theoretical physics through the lens of metaphysics. In the case of the all-natural universal meaning that had the longest wind-up—the optimization algorithm—I recognized that evidence for such a claim *if this was the only universe* was not only insufficient, it was also countered by the way reality actually works, not moving toward an optimum, but in accordance with the evolution of the wavefunction, and was tabled until we began discussing the multiverse. It is not because of our personal beliefs or because there is greater evidence for personal, natural-universal, or supernatural-universal meaning that we initially promote one hypothesis over the other, it is the rationale presented in each case. Nowhere in the hypothesis that meaning is inherent in existence were claims made that violate the laws of physics; instead, claims were simply made that there are other models, many of which depend on the existence of parallel universes in a multiverse, that also describe our current conception of purpose. Supernatural metaphysics gladly break the laws of physics, and from the perspective of a religious adherent, this is part of their charm, an added component of meaning, but a failing from the point of view of building an *explanation*—a self-perpetuating set of low-variability knowledge.

Supernatural metaphysics are bad explanations (in the Deutschian or Popperian sense outlined above) not because they lack evidence (although there is a great lack of evidence) but because they are highly variable. In the modern West, it is popular to believe in one of the monotheistic Gods of Abraham and to be an atheist to all other gods, including the Greco-Roman gods previously all the rage of the ancient Western World, but now relegated to myth. A strong case can be made that supernatural explanations are variable around political power.

Personal explanations for meaning are far less variable than supernatural meaning. Personal meaning settles on meaning that comes from optimizations in the experiences of well-being over those of unsatisfactoriness. While this meaning can be attained by a variety of activities, it explains meaning in an invariant way... focus consciousness, find meaning. Even without free will or an independent self, paying attention to the present contents and context of consciousness offers an unadulterated laboratory to first understand and eventually manipulate traits of loving-kindness, acceptance, and emotional control. It is easy to find meaning in this practice of self-improvement through paying attention to how your mind works or, in the case of optimizing your interrelatedness to existence or flow, by existing in a state of improvement at an important cognitive task.

Personal explanations for meaning have started to gain evidentiary support as studies on meditation practitioners and those that achieve flow states through long bouts of designed practice show signs of improved mental and physical health. However, separating experience into a domain separate from existence for the sake of solopsising meaning weakens the explanation, as we have seen by assuming that time flows, since that is our experience of it. Our felt experience, our subjective consciousness, may be as simple as informational processing at the level of the brain, in which case we would say that consciousness behaves classically and there is nothing more fundamental to experience than the existence of some sort of neural network. If that is the case, we should be able to recreate consciousness in classical computers and the hard problem of virtualizing consciousness is a technological leap we can make in this universe without reference to many worlds' parallelism. It would be surprising that of the many mysteries still open to scientific speculation: the nature of consciousness, the non-locality of space, the frozen flow of time, and the comprehensibility and apparent computational nature of existence, that the one that remains the most distanced to our understanding—namely consciousness—would be classical in nature. It is far more likely that, just as the dynamism of interference between near-parallel universes *feels like* the flow of time, so too is *feeling like anything* dependent on the interference of the multiverse we live in. If this multiverse and its informational processing of a parallel-psychic distribution of many minds is behind personal experience, it is reasonable to assume that it is also an important part of the optimization of said personal experience or meaning.

Science, Skepticism, and Woo

Migration between skepticism, cynicism, speculation, and woo for a mixed metaphysical and physics book on the meaning of life has been the most consistent and contentious source of edition. Science is humanity's best approach to comprehending existence, and I revere the scientists that have worked on their corner of existence to shine the light of their knowledge into the darkness. The awe I take in the description or depiction of the natural world is exponentially magnified by even my layman's understanding of it. Individual valuation is largely based in appreciation, and there is nothing I appreciate (or value) more than our human endeavor to gain knowledge of existence.

The danger I hope I have avoided is that of weakening science through speculation. The ditch opposite supernatural meaning is pseudo-natural meaning, a meaning flung far into the weeds of the latest misunderstanding of theoretical science. There will be those that will claim that this book has grazed or sunk into the ditch of pseudo-natural meaning and while their criticisms are taken seriously, I have made every attempt to separate the science from my speculation on what scientific discoveries might *mean for meaning*. I have worked hard to eliminate *woo*—magical claims drawn from a misunderstanding of nonintuitive science—the most diabolical monster pseudoscientists like those responsible for "The Secret" or "The Laws of Attraction" have wrought on mass markets still subject to snake oil salespeople. Instead, I have transparently proffered metaphysical frames of theoretical science, a way of thinking about reality described by science that challenges our parochial view of existence as just matter and energy. The metaphysical question of "what is the fundamental entity of existence" has perspective answers in modern science and a review of the informational hologram or the computational parallel-processing quantum computer offer different types of meaningfulness than the material "universe of record."

Our intuitions are not a good guide for the interplay between existence and experience or the meaningfulness of our lives. In order to overcome our faulty intuitions, I have mixed and tossed speculation and science, metaphysics and physics, spirituality and skepticism and not always in equal parts. The intuitions that needed tweaked and that guided our discovery include the experience of the flow of time and the intransigence of the hard problem of consciousness. Our ideas of each of these problems are caused by our most persistent, most universal agoraphobia, the fact that we cannot objectively observe ourselves outside the universe... for if we could take an "edge-on" view across near parallel universes, we would notice that both the present "moment" and our present "qualia" are actually in interference across a distribution of near-parallel universes. Neither the quantum nature of time nor the many minds interpretation of consciousness is my speculation, but together they offer an explanation that alters the most personal form of meaning—mindful, loving awareness of experience—into a universal meaning, a prevalent-personal meaning in the limit across the parallel-psychic consciousness.

These are falsifiable claims that exist in their time in science. If holography, the multiverse, or any of the other scientific claims utilized here in frames of metaphysics are proven wrong, or when our knowledge improves and these models are engulfed in a more complete one, than the meaning in the multiverse argued for here will either need to evolve or be discarded. While these hypothesized frameworks are beneficial in changing the way we view the universe and the meaningfulness of our lives to the cosmos, the science is still at the center. Our attachment has to be to explanations that are rigorously debated and, at some level of technology in the future, capable of being observed. If we can continue to make gains in more complex spacetime constructions and transform relativistic physics into quantum physics through the use of holographic processes, we will be able to state that the translation of information on the surface is constructed into material in the bulk through holographic means. As we come to understand the computational abilities and mechanisms of quantum computers, our ability to interrogate near-parallel universes will become more robust and explanatory. Furthermore, if we start to delve into a serious investigation of the quantum computational constructor of consciousness in an effort to solve the hard problem of virtualizing consciousness, we will be able to sort out if the interference of many minds is the neutral interlocutor that connects experience to existence.

For today's scientific Gantt chart, the above experimental plan represents a grandiose direction, but the near future of scientific progress is normally undershot by even our most futuristic minds and none of the above model validation tests is beyond the pale in the coming decades. Our commitment to these scientific, technological, engineering, and mathematical (STEM) *works* is not only posterity's most likely venue to remain gainfully employed, but it also underscores a meaningful path humanity can take that has a win-win positive-sum outcome, helping us solve the inevitable and often existential problems that only advances in knowledge can overcome. Along with the fact that the universe is "turned-on" where you sit, this whirling dervish of a virtuous cycle of problem solving is novel to intelligent life like us, and is capable of manipulating local and even intergalactic regions of the universe away from the advances of entropy and toward the more creative, aesthetic, sophisticated, and useful. The advances of a civilization in resourcing energy, computational complexity, spatial range, and subatomic manipulations are also advances *for the multiverse* that are separate and incomprehensible to *just* the laws of physics. The deep learning experiment run over eons may be agile, with each advance being a minimally viable product of its own meaningfulness and importance, or it may have an ultimate aim, which we at the "beginning of the infinity" cannot imagine, but something that requires our constant efforts against the present day existential filters that separate our posterity from this ultimate science experiment. We need to keep solving problems, our ability to do so by devising complexity out of randomness is unique and meaningful in the multiverse, and so, maybe in the far-flung future, will this meet a need of a fundamental nature.

The Morality of Meaning in the Multiverse

Agency to navigate to the most optimal consequences for human well-being is one of the cornerstones of our modern jurisprudence. The humanist values of the West, while not always manifest in its actions, regard humans as the masters of their own destiny, one that novelly trumpets the well-being side of the equation (life, liberty, and the pursuit of happiness), while writing laws to dissuade our inhumanity. Humanism, as part of the canon of the Enlightenment[126], alights a path by which our well-being can be codified in moral principles that do not require sacrifice to the needy whims of the god(s), but instead only ask us to reason through how it would be for other humans with other biology and other experiences. The power of our stories, of lovingkind living for a while in another person's shoes, becomes the key to unlocking the precepts of a whole moral system. Our humanism comes from understanding our own experiences, the best ways to alleviate suffering and optimize well-being for ourselves and from directing our loving-kindness out to others with different experiences that set their worldview askew (but approachable with shared imagination) to our own.

The arc of justice traced out by humanistic morality and the liberal democratic order have improved the lives of all modern humans, slowly turning us away from the wrong answers of biology-based bigotry and tribalism to a more open and global society. The moral order that was overturned by the slow progress of human liberty was guided by the universal moral compass of God(s). From what we can tell from most versions of the creator of the universe, he is not a humanist. From Zeus to Yahweh, our supernatural deities are petulant, conditional lovers of their creation. Their morality is simple: it is their way or the highway. Enforcement of the moral code of God(s) is, inexplicably, left to fallible mortals resulting in more tribalism and less error correction. Historically, only those lucky enough to claim a position in the prophet's hierarchy saw improvements in their well-being... for the rest, it was better luck in the next life.

The continued evolution of our *intersubjective reality* from religion to humanism was not agreed to in committee but is incentivized by the outcomes—the consequences—foreseeable from each. Furthermore, the secularization of belief is not the end of the changing landscape challenging humanity's collective moral intuitions. In *Homo Deus*, Yuval Noah Harari argues that many of the scientific advances discussed in this book like quantum computing and artificial intelligence have already germinated a new *intersubjective reality* of *dataism*[127]. Especially as privacy is concerned, dataism turns its back on the consequences to human well-being fundamental to the intersubjective reality of humanism and returns controlling interest for our cooperation back to a more "universal" entity—the ubiquitous "cloud." The incentives to human well-being in the short term being a life of custom experiences and the smart automation of most tasks but in the long-term being supplication of control to more complex algorithms capable, in theory, of solving objectively for our more global, chronic quandaries like climate change and nuclear proliferation. Harari's claim is that the enlightenment traded meaning for power; he makes the Huxleyan prediction that the disintegration of humanistic values will see humanity give up its power for comfort and improved capability.

The promise of the collaborative intersubjective reality of dataism and its spin off morality is that it will do a much better job by taking a more objective approach to triaging the problems and resourcing the solutions for global, chronic issues like climate change, nuclear proliferation, and global pandemics. These issues do seem to require updates to our humanist ethics because they are both global and chronic in nature. Global issues have been solved before, granted with great loss of life and destruction of resources, by the morally necessary rise of the Allied Powers to defeat the Axis in World War II. A valid counterfactual that war could have been avoided with a more objective peace to end World War I or with more objective data on militarization of Germany and Japan was even argued at the time by the prescient MP, Winston Churchill. More than humanism's failures in global moral quandaries, our ineptitude in countering problems that are chronic in nature demands improvements like those that are evolving out of the morality of dataism. Humanism struggles in the maintenance of the public good where our tendency, like when President Donald Trump skulked out of the Paris Climate Accords, is to let others expend the resources, while all reap the benefits. This is the "Prisoner's Dilemma" we are all in with respect to the existential threats that we face and need an objective, data-driven approach to overcome our past myopia.

This "Brave New World" is far from brave and while it trades an unsophisticated master for a high-tech one, it is likely just as naive. At its outset, the rise of dataism has been disorientingly negative. Algorithms utilized by Facebook, Twitter, and YouTube have promoted outrage, leading to a rise in disinformation without relevant accountability or editing, a breakdown in reasoned discourse that exacerbates partisan political tribalism, the rise of trolling-for-fun culture, and these tactics' effective use by third-world tyrants against first-world democracies. At a more individual level, bullying, identity theft, pornography, and the rise of online racism further erode our capability for humanity.

Even with a greater grant of data from a smart Internet of Things (IoT) than is currently predicted, the ability of our basic climate algorithms to act on data to conserve, innovate, and ultimately reduce carbon in the atmosphere is not likely within the window needed to avoid a 2.5 degrees Celsius rise in global temperatures. While our technological future will continue to push on and greater sensing and programmable control of all energy sources and sinks is a necessary component to reach a steady state, our ability to pass through the existential filter presented by global, chronic environmental and health problems will require that our humanist values be supplemented with and persuaded by meaning from the natural world.

As we have seen, a universal morality can be fraught with inactivity, wistfulness, and the relinquishing of power. The god(s) eventually stepping in to save the good and punish the bad offers the win-win ending to the sacrificing faithful. An all-natural universal morality must jettison this claim on the benevolence of even the most ingratiating natural entity—the superconscious quantum computer—since it is still more likely that we will have to start the technical evolution of such a machine before reaping its benefits. As to the rest of the meaningfulness on offer from the multiverse, the most we can expect is loving persuasion toward mindful optimal experience or flowing interrelatedness with existence, both requiring effort and advances in our knowledge. The moral change on offer from a persuasive, purposeful existence is not one of agency, but instead of inclusion; where humanism focused on our individual and often tribal well-being, multiversal morality takes as its objects the well-being of consciousness and our posterity-promoting problem-solving works.

Each of the objects of multiversal morality: the illumination of conscious states of well-being, including the solution to the hard problem of virtualizing consciousness, and the resourcing and innovation of solutions to existential problems, supplement the humanistic values of liberty, freedom, and justice with a necessary injection of perspective. The moral obligations of our ecological projects and spiritual development are pressing because of the meaningfulness where existence smears with experience. When we recognize there is no center and begin *being consciousness*, illuminating our little corner of the world, we recognize we are not distinct, but interrelated, a superposition of existence-experience, at times mindful of the now, and in other moments, creating knowledge and solving problems, inventing our shared future.

Multiversal morality is more proactive than other universal moralities and has a broader perspective than other personal moralities. It challenges us to build out the consequences not only to humans living today but also our posterity and the project of our survival and progress. Multiversal morality recognizes that humanity has a few unique qualities that should be perpetuated and that solutions to our survival, both on Earth and within our cosmic neighborhood, are within our current, and will be within our eventual means and knowledge. Our effort to reverse the worst effects of climate change—through smarter grids, algorithmically-controlled automated traffic, renewable energy sources, and incentives for public and private conservation and innovation—is the first works project to truly challenge our global scientific capacity and the persuasiveness of saving posterity within the multiversal morality. Further solutions to maintain vaccinated immunity, monitor the conditions that might incubate a rise of 'super-bugs,' and rapidly manufacture vaccines will have to run in parallel with geopolitical efforts to reduce nuclear weapons and promote the rise of science, liberty, and freedom for the entirety of the global community. Problems are inevitable, but problems are also solvable.

Few of the aforementioned improvements in civilization's capacity will have as large an impact as our pursuit of consciousness. Debugging our subjective feelings on the spectrum from suffering to well-being to an extent where we can safely and benevolently reduce unsatisfactory experience and greatly improve on happiness, bliss, wisdom, contentment, and all other sorts of well-being should continue to persuade the individual mindful practitioner, the brain-based neuropharmacologist, and the quantum computer programmer alike. Building machines that align through conscious second-person understanding of *what it is like to be human* enables collaboration with superintelligent machines likely necessary in our pursuit of intergalactic range and resourcing and the quantum computational advances necessary to eventually pursue a reversal (if such a thing is possible) of cosmic expansion and our deep descendants' fulfillment of the universe's deep learning problem.

The human quest for meaning has been continuous and creative, discovering the nature of existence and answering whether there is a grand universal purpose, is fundamental to our species. We are curious and industrious, yearning at once for mystery and closure. The combination of spirituality, philosophy, and science quell our appetite for updated questions, probe at answers, and cycle back. When we have been capable of unifying, it has been in the solution of a problem. Our creations from liberal democracy to the Large Hadron Collider are inspiring in their distinction from our primary programming: the mass multiplication of our genetic material.

Yet, we have also created more than our share of problems and we face new threats. Our social conquest of earth comes with responsibility to avoid the many ways in which we can inextricably alter the planet and our society. Our pressure on biodiversity and climate is our new great challenge. Our memes evolve in response to this fight.

Intellect and the ability to manipulate some of the mechanisms of the Earth's algorithms can continue to take us to good solutions. But implementation of these solutions is only possible if the vision takes a multilevel approach and brings us together under a compelling message to all of society: that we must work together, through each of us finding our optimal experience and returning to retell all of society of its majesty and flaws.

Solving problems and creating feats of engineering is impossible without vision. A shared vision intrinsically inspires, it keeps us late at the whiteboard, gathered with a diverse and multidisciplinary group, willing ourselves to look at the problem in a new way. Synergies where two heads are more than the sum of their parts (and mixed metaphors work out) are more likely in the zone—a flow created from a process that doubles back to compliment the achievement of shared purpose with intrinsic value.

Multiversalism suggests a greater appreciation for nature, knowledge, and science, continued development of rites to enhance consciousness and flow, and the broadening of consequentialism to include impacts to posterity. Meaning in the modern world comes from actively noticing our moments of flourishing and latching onto them, nurturing them for ourselves and in the service of others; from developing wisdom in our relationship with other conscious entities; from diving into a deeper understanding of the world around us; and, if not supporting it with our own efforts, from advocating on behalf of those making a difference in solving existential threats and confronting the cynics that reduce solution efficacy. Multiversalism is inclusive of meaning where our unique talents, and those of other conscious or intelligent beings, are essential to some grander project *for* the multiverse; meaning that is graciously granted to us by a superconscious invention of a parallel universe; and meaning that is the same as personal meaning—dependent on illuminating our consciousness with profundity—but that is prevalent across the multiverse. The metaphysics and ethics of multiversalism bring together the wonders of the physical world of existence—a computational multiverse capable of programming persuasive optimizations interfering with space, matter, consciousness, and time—with our experience that is responsible for the landscape of well-being and suffering we base our morality and personal meaning upon. The dualism of setting our *selves* and conscious experience outside of existence served us when this was the only universe and explanations required consistency in this framework. However, our universal myopia coupled with our self-disregard for our conscious and constructor talents—fairly unique in the universe—has caused humanity to accept a position of nihilism that we cannot (and even ought not) collaborate with existence to make improvements for our survival which impact and enrich the computational multiverse. Multiversalism places us on a million-year trajectory of problem solving, a beginning of infinity for our understanding of existence and the optimization of experience, not just for humanity, but as a part of the deep learning algorithms and optimization routines running as the meaning of the multiverse.

Meaning in the Multiverse

We are part of an awesome universe. A place that at its most fundamental is not some small variety of stuff or energy, not even a tiny packet of information or the merest concept, but, instead the humble process whose instructions turn geometry into gravity, the resolutions of fields into matter, and the interference of parallel experiences into consciousness.

Processes intermingle to create reality and our felt experience of it. This is readily apparent when we consider the reality of frozen time and our experience of it flowing. All of the equations of relativity tell us time is wrapped up with space in a continuum, the proverbial sheet we are all riding "along," its wells creating gravity, its stretching (as if forcing another imperceptible dimension into our cozy perceivable ones) the result of the mysterious dark energy.

Yet time appears to flow. Our experience of the present is fleeting, it is always being lost to our memories or galloping ahead as we chase our plans for the future. Time, like all quantum mechanical processes, is best understood by invoking the many worlds interpretation of quantum mechanics or the multiverse. Time does not flow in our experience of the single universe but instead time's flow is our imperceptible travel through the stack of near-parallel universes.

Time is not the only process that depends on the computational parallelism of the multiverse. The multiverse was first invoked to explain what was happening when a single particle shot at a dual slit resolved into a wave-like interference pattern. Instead of just adhering to the formulaic Copenhagen interpretation's ignorance of explanation, a multiverse of shadow particles interfering from their positions in near-parallel universes reframed the solution. This superposition of universes made the wavefunction a universal process and their interference responsible for the material universe. The idealistic universe, fundamentally bits of information or qualia of consciousness, also depends on computational parallelism to smooth over the rough edges of this ontology.

Explaining the meaning we recognize from our appreciation of consciousness as an outcome of a process of existence is a necessary step not only in finding meaning in the multiverse, but also in aligning our intelligent machines to the well-being of conscious creatures. The computational nature of existence is partially comprehended in classically compiled virtual reality and could be simulated to perfect fidelity in quantum computers, but there is as of yet no wavefunction of the mind available to code into machines and virtualize experience. A new field of quantum computational neuroscience is likely necessary to augment the many minds interpretation of consciousness and experimentally uncover the process of interference between near-parallel wavefunctions of the mind. Quantum computational neuroscience is a discipline suited to scientifically determine if the fundamental nature of reality is conscious or computational, how experience and existence relate at a fundamental level, and if we can solve the hard problem of virtualizing consciousness for our intelligent machines before it is too late. Even assuming these amazing theoretical outcomes are realized in our scientific future, it is the practical spinoffs in our ability to engineer greater well-being, retool criminal minds with neuro-restorative justice, and understand the neurocorrelates of depression, anger, and fear that will propel quantum computational neuroscience in i5.0—the fifth "industrial" revolution—the rise of quantum computation.

Quantum computation is at the place now that classical computation was at when computers were called Adding Machines. Our use of quantum computation in cryptography only scratches the surface of its capability. Utilizing the parallelism inherent in the many worlds interpretation of quantum computation to unlock the secrets of how consciousness and its persuasive optimization interfere to promote meaningfulness and well-being offers a possible consilience of experience and existence. Furthermore, the study of quantum computational neuroscience unifies the two ways we are meaningful *to the* multiverse—developing knowledge *about* consciousness. We our turning our eyes of world onto our universal constructions and computations and trying to understand their meaning and the manipulations we can do now to improve them. It is not enough that we create intelligent machines to assume our role as the intellectual hegemon, but instead we must create them to be conscious so they can interact with care and kindness even beyond our galactic region.

The only ontology that reasonably explains and unifies existence and experience is one of quantum computational parallelism. Most of the aspects of the natural world and the landscape of the mind that are paradoxical when considered in a single universe are clarified when considering influence across parallel universes. Our experiences exist in continuum with many minds and this parallel-psychic wavefunction of the mind is the neutral interlocutor that makes it *like something* to be a collection of quantum particles. Like how the interference between near-parallel universe appears to be the flow of time, or how the process of a multiverse of interfering wavefunction evolutions appears to collapse into one outcome in our universe, our conscious experience exists as a distribution of more or less conscious entities across near-parallel universes. There is no distinction between the entities of existence—matter, energy, time, space, information—and those of experience—thoughts, consciousness, sensations—it is process all the way down.

Some of these processes are inherently meaningful, even before we dally into the metaphysics behind them. Most obviously, our ability to appreciate experience grants us with the *only* perspective that can produce love and compassion. Our mindfulness of subjective conscious *likeness* and its *felt* and *narrative* relation with those around us improve our interactions, society, and our posterity. Our moral standing is in relation to the consequences scored in terms of improved conscious states, namely: the production of well-being and the reduction of suffering. As we organize society, our internal liberation from self, grasping, and distancing is coupled with our narratives of loving-kindness in relation to one another and posterity and create liberal, scientific democracies. In these always improving societal systems of collaboration, progression, and creativity, our species will ultimately align to an intersubjective entity not selfishly humanistic nor one-sidedly scientific, but a more universal vision of a scientific description of consciousness and its dissemination into our machines and ultimately to program the processes of the multiversal quantum computer.

The only thing that stands between our present capabilities and a very utopian future where we rest having achieved the dissemination of intelligence and consciousness into the wider universe is the knowledge to do it. The problems, both scientific and societal, that stand in our way are inevitable; but they are solvable. Our ability to optimize our little corner of existence through the virtuous cycle of flow-producing deliberate practice is the progenitor of great art, sport, science, and any other human endeavor. It will take all of our collective efforts—we will be throwing complex systems against complex problems—so business and governmental resourcing, scientific invention, and societal alignment through art and music.

We are a special sort of biological scum floating around on an indistinct rock. Our potential to be the meaning *for* the multiverse, the continued growth of our knowledge and our subjective loving awareness, is underappreciated. We have not evolved to do this, but now that we are here, our vision for our species should not be so pessimistic to our capabilities or limited by our prescientific beliefs. Uncovering the mystery of how experience arises from existence is the cosmic egg to our multiversal chicken, for only as we start to use quantum computers to model parallel-psychic machine consciousness will we be able to unlock the best well-being for our near-parallel selves, our posterity, and a new superintelligent, superconscious computer that can take over running all of the processes with optimizing well-being for all sentient life. We will need an entity like this superconscious massively parallel multiversal quantum computer to solve our universal problem of the heat death and to reverse the current course of spacetime's accelerating expanse. However, even saving the universe from the heat death would pale in comparison to the instantiation of consciousness into the multiverse and the likely returned grace of optimized well-being, heaven on earth, from the newly superconscious near-parallel optimization process. Realizing our special conscious and epistemological meaning and motivating our actions to be more personally mindful and supportive of society's species-saving scientific efforts is how we start.

We are all bits of being, distinct and useful to the processes fundamental to the computational multiverse. Nurturing awareness or multidisciplinary intellect leads some to a realization of magnificent meaning. Still others achieve eudemonic happiness from creating soundscapes in the moment, delivering fiction in film, or playing a game at peak performance, creating a lifetime of lengthening optimum experience. Others work from the inside out, smearing what is true about experience onto every moment until those experiences attain the richness of purity; instead of flow, there is only context, a vessel where meaning is made manifest, our subjective consciousness where value arises.

Aligning to multiversal meaning is a marvelous and awe-inspiring journey. If you put your mind to it, each and every moment, material and information and process can thrill you. It can be artful and inspiring, maddening and desperate. We fall in and out of love with it all. It is inexplicable and incomplete, and wow is that intriguing. To be inspired is part of the path.

There is no need to fear the darkness of deep space or seek meaning in material. What humans lack in current cosmic range, computational power, or energy resourcing, we can learn; while our societal gracelessness-under-pressure, tribalism, and lack of unifying vision can be overcome by greater mindfulness. Our natural world is a complex quantum computer not only projecting parallel existence, but also persuading us to greater well-being until we arrive at an algorithmic optimum, experiencing the meaning in the multiverse.

Acknowledgements

AS AN AMATEUR NATURAL philosopher, I am indebted to all of the scientists and philosophers whose work and books have influenced me and the ideas that are in *Meaning in the Multiverse*. It is my profound hope that I have done the rigor of their work justice, for while this is largely a book of conjecture and hypothesis, there are scientists and engineers out there that have compellingly and with great precision explained existence in a way that is backed by reams of mathematical theory and experimental data. Some of these scientists include David Deutsch, Donald Hoffman, Ray Kurzweil, Max Tegmark, Seth Lloyd, Alan Lightman, Michio Kaku, Sam Harris, Richard Feynman, Stephen Hawking, Richard Penrose, Carl Sagan, Juan Martín Maldacena, Brian Greene, and the giants that came before them whose names stand testament on equations that demarcate the greatest leaps in human knowledge: Turing, Einstein, and Schrödinger.

The reinvigoration of my mindfulness practice and deepening appreciation for the loving awareness of consciousness is thanks to Sam Harris and his *Waking Up Course*. Sam's instructions are clear and offer a path to more durable and useful states of mindful experience. Furthermore, he has injected the app with profound lessons like *The Headless Way* by Richard Lang and *Contemplative Action* by David Whyte that elaborate on the noetic mysteries at the center of first-person subjective experience.

Friends who read early drafts of first-time writers' books are a special kind of masochist and I am indebted to my long-time achievabilibuddy Ian McDaniel for always giving me great notes

ACKNOWLEDGEMENTS

and even greater encouragement; to Mark Johnson for stretching my philosophical capabilities to see grace and love in the computational universe; and to Joe Badal for keeping it light. The book is much improved thanks to my editor from The Manuscript Doctor, Emma Farnsworth. Finally, self-publishing was only possible while working from home during the coronavirus thanks to Stephanie Chandler and the wonderful group at the Non-Fiction Authors Association who not only supply valuable information, but who also manage to motivate in 2D.

My blended family is incredibly supportive while offering a unique perspective and no-nonsense thoughtfulness on any topic available for loud and complex debate. Huda and Dan have been (mostly-) willing participants in the majority of my marathon all night discussions and have been even more generous sharing the love and kindness of their daughters, Layla and Naima, who are at the center of my heart and have taught me something every day I am with them. Osama has a talent for everything, and his projects are a thrill to follow and learn from him the intricacies of hobbies of a true autodidact. Naba is the most unquestioning and supportive of my writing, sure that it will jump off the page because of the goodness she wants for me and the family; her support only greater for her husband John who should be the Mayor of Boise as he brings "the heat of a brand new day," every day and for her smart and sweet young children Nicalona and Keppa. Hadiel reminds me so much of her mom in her fight and in my desire to impress her with my wit and wisdom, yet she shares a comfort in the simple things that keeps her and her many friends, grounded and secure in the space she makes for them and their shared struggles.

My mom has always taken an interest in my advocations and hobbies, whether baseball, service, hiking, or metaphysics. She is an unstoppable force behind my brother, Ryan, and I. She read the most discombobulated drafts and asked for explanations in a way that protected my ego (illusionary though it is) and conveyed the work ahead. That my mom and dad showered me with love and provided an expansive and fun childhood with a younger brother that was up for anything is the best break a kid could ask for, thankfully I didn't squander it.

Samira is more than the love of my life, my best friend, and my Executive Director—she is mythic. I've always thought so. When you hear about her life, including now, you mistake it for a story told by a wizened and wise storyteller, complete with morals of service and persistence, love and humor, and a long line of matriarchs passionate about family and living life to its fullest. The romanticism of the Iraq of her childhood was quenched and hardened by her innovativeness as one of the first Middle-Eastern women engineers in semiconductor R&D and this sharp yet delicate instrument was put to its best use in the service of forcibly displaced women in the non-profit service organization she founded. She is my all-natural muse.

About The Author

JUSTIN HARNISH IS A writer, engineer, and non-profit founder. He is the author of the book *Dance to Spawn A Galaxy* and has collaborated on numerous patents and papers in semiconductor engineering and data science.

Justin is a free-range thinker, a futurist, mindfulness practitioner, and a speculative Natural Philosopher. He studied Chemical Engineering and Philosophy at Montana State University. As part of his undergraduate degree work study, he wrote for the Wallace Stegner Endowed Chair website starting a lifelong love for the stories of place and the evolution of people through their lives.

18 years in the School of Life, Work, and Family educated Justin in the diverse pursuits required of examined life living. Coupling the science of his engineering education and early career with the philosophy of strategic design and management, Justin found a path in 2012 to couple them, becoming the Technology Program Manager for the semiconductor manufacturer he works for. Revitalizing the preeminence of the classic Natural Philosopher -- an awe-inspired creative hypothesizer, futurist on one hand, a pragmatic, methodical skeptic on the other -- is Justin's life work.

Justin's commitment to thought, spirituality, and our shared story is exemplified in his founding of the Jung Society Think Tank. The Jung Society Think Tank is a "book club without a book," a place where those we might (tongue-in-cheek) call "amateur thinkers" join together to discuss one of the group's extracurricular research. In the twelve-month history of the JSTT, its members have covered topics ranging from the Neuroscience of Memory to the Sacred Life Of Bees: The Return of the Divine Feminine.

Utilizing the life optimizing strategies outlined in Meaning in the Multiverse, Justin helps people detail their passions and align their life and professions to achieve their greater purpose.

Justin serves his community as the Creative and Development Director of Women of the World (womenofworld.org), a non-profit women refugee service organization he founded with his lovely wife Samira Harnish. Women of the World offers custom critical service and capacity building to displaced women resettling in Salt Lake City, where Justin works and lives. Samira was recently awarded the Americas Nansen Award from the United Nations High Commissioner for Refugees for excellence in refugee service.

Justin loves his blended family including his wife Samira, Samira's 4 children, 3 granddaughters, 1 grandson, and his Silky Terrier (formerly known as) Prince!

Web: https://justinaharnish.com
Twitter: @justinaharnish

Endnotes

"Ain't it just like the night,

to play tricks when you're trying to be so quiet?

We sit here stranded,

though we're all trying our best to deny it...

Inside the museums,

infinity goes up on trial.

Voices echo

this is what salvation must be like

after awhile."

Bob Dylan, "Visions of Johanna"

1 The danger in making metaphysical claims from experience is noted, however, the flow of time is a universal experience and more of a failure of our language than even a failure of intuition or an experience making claims on the physical state of existence. We say that time passes because we have clocks that mightily tick off the 100th of a second in many cases, but what has really passed? Did time flow by us like a gust of wind or through us like a neutrino? The flow of time is a speculation that has dutifully passed as fact up through the ages until relativistic spacetime physics apparently stopped it in its tracks. However, conjecture by physicists like David Deutsch will show that time is not "frozen" as just another dimension in the spacetime continuum, not flowing, but that time is the tunneling of our entire universe as a single quantum distinction changes our universal wavefunction. If our ancestors had known of the multiverse, they would have spoken not of the flow of time but the tunneling to parallel universes.

2 Sagan, Carl. Fall 1987. "The Burden of Skepticism." *Skeptical Inquirer.*

3 The "replacement" suggested here is solely in the tradition of improved scientific concepts furthering those of old like how classical physics remains for a subset of conditions but is subsumed by quantum physics as the most accurate approach.

4 Hesse, Hermann. *The Glass Bead Game.* New York, NY: Stellar Classics, 2013.

5 Adams, Douglas. *The Hitchhiker's Guide to the Galaxy.* London: Pan Macmillan, 2016.

6 Camus, Albert, and Justin O'Brien. *The Myth of Sisyphus.* New York, NY: Vintage International, 2018.

7 Ibid.

8 While group selection mechanisms still require more study, the social conquest of earth by humans is the greatest increase in intelligence that we are aware of and likely holds a clue to the requirements for improving invented intelligence. The causal linkage between language, migration of game, and, interesting while lesser studied contributions from psychedelic compounds gathered along the way marks the rise to prominence of our species and of the early

stories we told ourselves that gave those lives meaning.

9   Harari, Yuval N. *Homo Deus: a Brief History of Tomorrow*. New York, NY: Harper Perennial, 2018. Harari claims humanistic control over our destiny overtook religious meaning especially in the Western world starting during the Enlightenment. Now our technological world is leading to another construct guiding our meaning away from humanism toward dataism.

10   The bottleneck is not technological, instead we lack good collaborative and decision-making methods to counter global problems. Even when we look at the data, we do so behind a veil of national interest. Our history of national success blinds us to the likelihood that most existential problems must be solved at a different level and with global partners.

The problems are looming, global, and existential; however, they are all solvable. Our scientific, technical, and computational creativity is up to the task. What remains to be seen is if we can overcome our bickering to triage and define issues and collaborate on their solutions. This will require a meaning that is capable of aligning us to something larger, a meaning based in our future or in meaning we extract from the universe itself. If we can align ourselves with a more species oriented motivational meaning, a long future of transcendent human well-being is very likely.

11   Harari, Yuval N. *Homo Deus: a Brief History of Tomorrow*. New York, NY: Harper Perennial, 2018.

12   Bostrom, Nick. *Superintelligence: Paths, Dangers, Strategies*. Oxford: Oxford University Press, 2014.

13   *The Pleasure of Finding Things Out*. BBC, 1981.

14   Damasio, Antonio R. *Self Comes to Mind: Constructing the Conscious Brain*. New York, NY: Vintage Books, 2012. These concepts from Damasio are experientially available to even the most inexperienced meditator: the body scan resolving itself into a cloud of sensations; feelings of disquiet, anxiety, and disgust on one hand and bliss, calm, and contentment; and the ever present stream of consciousness storyteller narrating even your most inane thoughts.

15   Jackson, Frank (1982). "Epiphenomenal Qualia". Philosophical Quarterly. 32 (127): 127–136.

16   Don't take my word for it, this is a great object of meditation. Being able to observe the felt nature of experience <u>and</u> not identify with it in the moment requires constant practice but is the well-spring for positive character traits like a reduction in anxiety and increased contentment.

17   Harris, Sam. *Waking Up: A Guide to Spirituality Without Religion*. New York: Simon & Schuster, 2014, pp. 146-147.

18   Ibid.

19   Hofstadter, Douglas R. *I Am a Strange Loop*. New York : London: BasicBooks ; Perseus Running [distributor], 2007, p.363.

20   Headlessness is one of the most obvious but profound concepts I have come across in the description of conscious experience. In *On Having No Head,* we are charged to "consciously be what we are- capacity for things. Reuniting them with the [source] the infinity that lies "this" side of it." Much more where that comes from, an essential read. Harding, D. E. *On Having No Head: Zen and the Rediscovery of the Obvious*. London: The Shollond Trust, 2014.

21   Harris, Sam. *Waking Up: A Guide to Spirituality Without Religion*. New York: Simon & Schuster, 2014, pp. 154-157.

22   Sagan, Carl, Ann Druyan, Adrian Malone, Steven Soter, Gregory Andofer, and Rob McCain. *Cosmos*. New York, NY: Random House Inc, 2013.

23   Einstein, Albert. *Relativity: the Special and General Theory*. London, UK: New Academic Science, an imprint of New Age International (UK) Ltd., 2019.

24   Smolin, Lee. *Three Roads To Quantum Gravity*. S.I: Basic Books, 2017, p. 53.

25   Deutsch, David. *The Fabric of Reality*. London: Penguin, 1998, p.243.

26   A recognition that at some level much "bigger" than the fabric of the universe, we are integrated with other "stuff" is one of the most profound realities for us to wrap our minds around. It doesn't mean you can pass through walls, but it does compellingly use scientific realities to show that we are integrated with the universe, not separately moving around "on" it. Instead, we are made, most fundamentally, of the universe, and nothing we physically do or mentally think is separate

from what we normally think of as existence.

Non-locality further complicates the idea of space, where defining a position on the spacetime continuum is "like placing a flag in the ocean." Where the space you are occupying is positioned at any time is not as straight-forward as it would seem from our macroscopic perspective. At the fundamental levels of the makeup of material, there is more unity than division, more network than node.

27   Our species's physical size and dexterity in intergalactic travel are some of the oldest and most meaningless measures of our value in the universe and say little about the likelihood of universal meaning. Far more tangible traits to our place in the purpose of the cosmos is our ability to manipulate classical and especially quantum computation in developing knowledge and constructing universal machines and the fact that we have a conscious experience of the universe. A more meaningful universal conception is one that assumes that intelligence and conscious experience are the most important distinctions in processes as fundamental to the universe's makeup and meaning as the current known Laws of Physics.

28   Lightman, Alan P., and Grover Gardner. *Einstein's Dreams*. Ashland, OR: Warner Books, 2015.

29   Goleman, Daniel, and Richard J. Davidson. *Altered Traits Science Reveals, How Meditation Changes Your Mind, Brain, and Body*. NY, NY: Avery, an imprint of Penguin Random House LLC, 2018.

30   Deutsch, David. *The Fabric of Reality*. London: Penguin, 1998.

31   The most famous laboratory setup for making a quantum distinction is the dual-slit experiment where interference between particles in near-parallel universes acting as defined by the wavefunction create different patterns on the film screen behind the slit apparatus and in so doing move the present along slightly. But any quantum distinction, even one out in deep space, unobserved and unrecorded, will do.

32   This is the method used in floating gate NAND Flash memory. This type of non-volatile (the bits do not require constant electrical refresh and are maintained when the power is shut off) memory was being manufactured by the most advanced memory manufacturers around the world in about 2010.

33   *The Matrix.* DVD. Las Angeles, CA: Warner Home Video, 1999.

34   Dennett, Daniel Clement, and Paul Weiner. *Consciousness Explained.* New York, NY: Little, Brown and Company, 2007.

35   The Large Hadron Collider or LHC is humanity's most complex scientific instrument. It is a twenty-seven kilometer track buried below ground at the border of France and Switzerland made of superconducting magnets that accelerate particles to speeds near the speed of light. One billion collisions are recorded each second.

36   Al-Khalili, Jim. *Quantum: A Guide for the Perplexed.* London, UK: Weidenfeld & Nicolson, 2012.

37   "Standard Model." Wikipedia. Wikimedia Foundation, August 2, 2020. https://en.wikipedia.org/wiki/Standard_Model.

38   Al-Khalili, Jim. *Quantum: A Guide for the Perplexed.* London, UK: Weidenfeld & Nicolson, 2012, p. 66.

39   Vedral, Vlatko. "Living in a Quantum World." *Scientific American* 304, no. 6 (2011): 38–43. https://doi.org/10.1038/scientificamerican0611-38.

40   There is not a great word for the interaction of universes at the moment of wavefunction collapse. Generation is not correct because there are many near-parallel universes that are fungible with the wavefunction in our "universe" and so can interact and evolve with minor distinctions from our universal wavefunction. So collapse in our universe will have a positive spin electron, in our near-parallel, near-"twin" universe, there will be a negative spin, both in accordance to the Schrödinger equation and not necessarily generating anything new.

41   Freese, Katherine. "The Dark Side of the Universe." *PUBLIC LECTURE SERIES | Perimeter Institute.* Lecture presented at the PUBLIC LECTURE SERIES | Perimeter Institute, 2018.

42   "Theoretical Physics: The Origins of Space and Time." Nature News. Nature Publishing Group. Accessed August 8, 2020. http://www.nature.com/news/theoretical-physics-the-origins-of-space-and-time-1.13613.

43   One of the major difficulties in the deductive logic of scientific discovery is the problem of priors known as the "turkey problem." A turkey born on Boxing Day has increasing confidence in his prior

conviction that the farmer is a good guy, a provider of sustenance, right up to Thanksgiving morning. Our observations of deep space and time heavily weight the likelihood that the universe will continue to act as it always has. This is a very good assumption since it would be very unlikely that the universe would change course over the short duration since we have started doing science. However, the symmetries that show up in our mathematical models do allow for the universe to "switch" from expansion to contraction, however, we have no prior for this and can only speculate on its mechanism.

44   Carroll, Sean M. *The Big Picture: on the Origins of Life, Meaning, and the Universe Itself*. New York, NY: Dutton, 2017.

45   Stroll, A., and R. H. Popkin. *Philosophy Made Simple*. Hoboken, NJ: Taylor and Francis, 2012, p.132.

46   Davies, Paul, and Niels Henrik Gregersen. *Information and the Nature of Reality: from Physics to Metaphysics*. Cambridge: Cambridge University Press, 2014.

47   Harris, Annaka. *Conscious*. New York, NY: HarperCollins, 2019.

48   Gleick, James. *The Information: a History, a Theory, a Flood*. New York, NY: Pantheon Books, 2011.

49   Davies, Paul, and Niels Henrik Gregersen. *Information and the Nature of Reality: from Physics to Metaphysics*. Cambridge: Cambridge University Press, 2014, p.148.

50   Davies, Paul, and Niels Henrik Gregersen. *Information and the Nature of Reality: from Physics to Metaphysics*. Cambridge: Cambridge University Press, 2014, p.146.

51   Gleick, James. *The Information: a History, a Theory, a Flood*. New York, NY: Pantheon Books, 2011.

52   Musser, George. "Where Is Here?" *Scientific American* 313, no. 5 (2015): 70–73. https://doi.org/10.1038/scientificamerican1115-70.

53   Greene, Brian. *The Hidden Reality: Parallel Universes and the Deep Laws of the Cosmos*. New York, NY: Vintage Books, 2013, p. 400

54   Moyer, Michael. "IS SPACE DIGITAL?" *Scientific American* 306, no. 2 (2012): 30-37. Accessed August 8, 2020. www.jstor.org/stable/26014198.

55   Greene, Brian. *The Hidden Reality: Parallel Universes and the*

Deep Laws of the Cosmos. New York, NY: Vintage Books, 2013, p. 400

56   It is hard to appreciate the size of this exponent. If universal information is stored on the surface of the universe in Planck length-sized surfaces, you are slicing up one of the largest things (the surface of the observable universe) by the smallest unit of measurement. It is a trillion trillion trillion trillion trillion trillion trillion trillion trillion trillion bits of information. By comparison, all of the leptons and bosons combined are only believed to count to about 10^{80}.

57   Gleick, James. *The Information: a History, a Theory, a Flood.* New York, NY: Pantheon Books, 2011.

58   The idea of a conscious entity of which we are just component or qualia is no longer a theory based in theology. Donald Hoffman's work on Conscious Realism is groundbreaking in that it comes from his well-founded and modeled claim that what we see as reality is just an interface utilized by our perception to display those items most relevant to our evolutionary fitness and to eliminate everything else. From this Interface Theory of Perception, Hoffman moves to a universe that is not primarily physical but a dynamic relationship between conscious agents and their felt experience of the interface. The Qualia of God Meaning and Pansychic Cog-in-the-Machine Meaning are both logical outgrowths from a universe based in conscious realism. Hoffman, Donald. *Case Against Reality: Why Evolution Hid the Truth from Our Eyes.* New York, NY: W W Norton, 2020.

59   Carroll, Sean M. *The Big Picture: on the Origins of Life, Meaning, and the Universe Itself.* New York, NY: Dutton, 2017.

60   Mesle, C. Robert. *Process-Relational Philosophy an Introduction to Alfred North Whitehead.* West Conshohocken, PA: Templeton Pr., 2009, p.8.

61   Lin, Derek. *Tao Te Ching: Annotated & Explained.* Woodstock, VT: Skylight Paths Publishing, 2006.

62   Stroll, A., and R. H. Popkin. *Philosophy Made Simple.* Hoboken, NJ: Taylor and Francis, 2012, p.127.

63   Mesle, C. Robert. *Process-Relational Philosophy an Introduction to Alfred North Whitehead.* West Conshohocken, PA: Templeton Pr., 2009.

64   Hariharan, P. *Basics of Holography.* Cambridge: Cambridge

University Press, 2002.

65   Greene, Brian. *The Hidden Reality: Parallel Universes and the Deep Laws of the Cosmos*. New York, NY: Vintage Books, 2013, p.543.

66   Ibid.

67   Freiberger, Marianne. "The Illusory Universe." plus.maths.org, March 13, 2020. https://plus.maths.org/content/illusory-universe.

68   Cowen, Ron. "Simulations Back up Theory That Universe Is a Hologram." *Nature*, 2013. https://doi.org/10.1038/nature.2013.14328.

69   Hesla, Leah. "Searching for the Holographic Universe." symmetry magazine, 2014. http://www.symmetrymagazine.org/article/april-2014/searching-for-the-holographic-universe.   The experiment that will decipher holographic noise involves measuring the relative positions of large mirrors with increasing precision. Since the holographic universe is discrete (a web one Planck's length on each side), information about the positions of the two mirrors is finite, so the researchers should ultimately hit a limit in their ability to resolve the mirrors' respective positions.

70   Freiberger, Marianne. "The Illusory Universe." plus.maths.org, March 13, 2020. https://plus.maths.org/content/illusory-universe.

71   Smolin, Lee. *Three Roads To Quantum Gravity*. S.I: Basic Books, 2017, p.170.

72   It is important that we distinguish the classical computer from the quantum computer. While both are universal machines, that is, they perform substrate independent computations in accordance with the fundamental laws of epistemology and information theory, quantum computation utilizes the fundamental constructor of quantum physics for its computations, while classical computation uses a classical subset.

73   Holt, Jim. *Why Does the World Exist?: An Existential Detective Story*. London: Profile, 2013.

74   Deutsch, David. *The Beginning of Infinity*. New York, NY: Penguin Books Ltd, 2011.

75   Ibid.

76   Hoffman, Donald. *Case Against Reality: Why Evolution Hid the Truth from Our Eyes*. New York, NY: W W Norton, 2020.

77   Pollan, Michael. *How to Change Your Mind: What the New Science of Psychedelics Teaches Us about Consciousness, Dying, Addiction,*

Depression, and Transcendence. New York, NY: Penguin Books, 2019.

78   Christian, Brian, and Tom Griffiths. *Algorithms to Live by: What Computers Can Teach Us about Solving Human Problems.* New York, NY: Henry Holt and Company, 2016, p.261.

79   Deutsch, David. *The Fabric of Reality.* London: Penguin, 1998, p. 46.

80   Ibid.

81   Greene, Brian. *The Hidden Reality: Parallel Universes and the Deep Laws of the Cosmos.* New York, NY: Vintage Books, 2013.

82   Given that there are estimated to be somewhere near 10^{80} elementary particles in the universe, the distinguishing rate would be very roughly 10^{-80} seconds—which is far faster than even Planck's time, the time it takes light in a vacuum to travel 1 Planck's length. Even this quick movement of light would be tunneling through ~10^{36} near-parallel universes in the briefest of moments available to relativistic physics.

83   Mesle, C. Robert. *Process-Relational Philosophy an Introduction to Alfred North Whitehead.* West Conshohocken, PA: Templeton Pr., 2009, p.108.

84   Nagel, T. (2002). *Concealment and exposure: And other essays.* New York, NY: Oxford University Press.

85   It is important to reiterate that this is all part of a computational universe and has transcended a quantum physical requirement in the information processor of any sentient being. The microtubules of Orchestrated Objective Reduction proposed by Penrose are not necessary just as silicon, copper, tungsten or any of the other elements used in semiconductor processor technologies are not required to have wires, transistors, or capacitors intrinsically in their makeup. The multiverse's parallel processing happens in brains as a result of the heretofore unknown interference of the wavefunction of the mind and no constructor beyond its parallel processing quantum computations is required.

86   Deutsch, David. *The Fabric of Reality.* London: Penguin, 1998, p. 165.

87   Ibid.

88   The near future utilization of quantum computers to

study the neuroscience of consciousness will be an exciting field of research that should help us understand how existence (especially informational processing components of existence like brains) creates consciousness at some fundamental level. Initially, I believe this will be modeled using classic neural networks coded for quantum computation or quantum computer-run VRs but will eventually need to consider the computational nature of the quantum computer to be the *constructor* of experience itself, outside the model. The rising power of quantum computation should hopefully ease the barrier to exiting the modeling stage to being in the constructor phase, where true, novel conscious experience can be created.

However, there is still another hurdle for these future quantum computational neuroscientists to overcome, one that'll likely be a far more astonishing leap to make at the level of technological sophistication of quantum computers in even the next one hundred years, which will be to nail down an explanatory equation for how consciousness evolves from its fundamental constructor of quantum computation into our felt experience.

Even if the parallel-psychic view of consciousness is not a valid explanation, there is one thing to push scientific research institutions for in order to further develop the science of the mind, and that is greater use of quantum computation in the study of neuroscience.

89   Csikszentmihalyi, Mihaly. *Flow: the Psychology of Optimal Experience*. New York, NY: Harper Row, 2009.

90   Ibid.

91   Ibid.

92   Ibid.

93   Ibid.

94   Colvin, Geoff. *Talent Is Overrated*. London: Penguin Books, 2008.

95   Stanford d.school. Accessed August 8, 2016. https://dschool.stanford.edu/.

96   Brown, Tim. *Change by Design*. New York, NY: HARPER COLLINS, 2009, pp. 148-150.

97   Schawbel, Dan. "Josh Kaufman: It Takes 20 Hours Not 10,000 Hours To Learn A Skill." Forbes. Forbes Magazine, December 17,

2013. https://www.forbes.com/sites/danschawbel/2013/05/30/josh-kaufman-it-takes-20-hours-not-10000-hours-to-learn-a-skill/.

98   Colvin, Geoff. *Talent Is Overrated*. London: Penguin Books, 2008, p.71-72.

99   Ibid.

100   McKeown, Greg. *Essentialism: The Disciplined Pursuit of Less*. London: Virgin Books, 2014, p.4.

101   "If it's not a *hell yes*, it's a no!"

102   Sinek, Simon. *Start with Why: How Great Leaders Inspire Everyone to Take Action*. London, England: Penguin Business, 2019.

103   Jagerson, Elisa. "Engineering the Ultimate Experience." *Association of Strategic Planning Annual Conference*. Lecture presented at the Association of Strategic Planning Annual Conference, 2014.

104   McKeown, Greg. *Essentialism: The Disciplined Pursuit of Less*. London: Virgin Books, 2014, p.4.

105   Logic models are widely used in non-profit development and strategy setting. There are numerous other types of strategy setting tools out there like the Balanced Scorecard—but what is critical is that there are measurements available for all activities and outcomes, whether they be completion based, quantitative, qualitative, tangible, intangible, explicit, implicit, direct, or line of sight. Contrary to popular perception and Dilbert, there is some real creativity involved in the engineering of metrics around goals.

106   Tolle, Eckhart. *The Power of Now: a Guide to Spiritual Enlightenment*. Sydney, NSW: Hachette Australia, 2018.

107   Langer, Ellen J. *Mindfulness*. New York, NY: LITERA MEDIA GROUP, 2020.

108   Zaleski, Philip, and Carol Zaleski. *Prayer: a History*. Boston, MA: Houghton Mifflin, 2006.

109   Everly, George S, and J. M Laiting. *A Clinical Guide to the Treatment of the Human Stress Response*. New York, NY: Springer, 2002.

110   Ibid

111   Kaplan, Aryeh. *Jewish Meditation: a Practical Guide*. New York, NY: Schocken Books, 1995, p. 40-41.

112   Campbell, Joseph. *The Hero with a Thousand Faces*. Novato, CA: New World Library, 2008.

113   *The Last Temptation of Christ*. United States: Universal, 1988.

114   Campbell, Joseph. *The Hero with a Thousand Faces*. Novato, CA: New World Library, 2008.

115   "Bhumisparsa." V&A Search the Collections. Accessed August 8, 2016. http://collections.vam.ac.uk/item/O82512/sculpture-sculpture-unknown/. Sakyamuni at point of enlightenment (Earth Witness): The Buddha made this gesture, known as Bhumiparsa just before his enlightenment to call the earth Goddess witness to his worthiness to become a Buddha. In response the earth shook and the evil demons of Mara who had been tormenting him fled in terror.

116   By far my favorite conception of enlightenment is one of intimacy with loving awareness, by Frank Ostaseski: "I prefer the word intimacy because it is an invitation to come closer, to fully embrace and lovingly engage with your life right where you are, rather than trying to move beyond it. It is a recognition that we already belong. To me, intimacy better expresses what I imagine enlightenment might actually feel like." Ostaseski, Frank, and Rachel Naomi Remen. *The Five Invitations: Discovering What Death Can Teach Us about Living Fully*. New York, NY: Flatiron Books, 2019.

117   His Holiness The Dalai Lama. *The Universe in a Single Atom: the Convergence of Science and Spirituality*. New York, NY: Three Rivers Press, 2005.

118   Campbell, Joseph. *The Hero with a Thousand Faces*. Novato, CA: New World Library, 2008.

119   Goleman, Daniel, and Richard J. Davidson. *Altered Traits Science Reveals, How Meditation Changes Your Mind, Brain, and Body*. NY, NY: Avery, an imprint of Penguin Random House LLC, 2018.

120   Ricard, Matthieu. "Neuroscience Reveals the Secrets of Meditation's Benefits." Scientific American. Scientific American, November 2014. https://www.scientificamerican.com/article/neuroscience-reveals-the-secrets-of-meditation-s-benefits/?redirect=1.

121   Lieff, Jon. "Meditation and the Brain 2013." Jon Lieff, MD, February 3, 2020. http://jonlieffmd.com/blog/

meditation-and-the-brain-2013.

122 Ibid.

123 Thanks to Sam Harris and the Waking Up App for his thoughtful and practical detailing of the modes of loving-kindness: compassion and sympathetic joy. Harris, Sam. Whole. Waking Up Course, 2018.

124 Ibid.

125 King, Stephen. *On Writing: A Memoir on the Craft*. N.Y., NY: Recorded Books, 2000.

126 Best delineated, with proof-positive data, by Pinker, Steven. *Enlightenment Now: The Case for Reason, Science, Humanism, and Progress*. New York: Viking, 2018.

127 Harari, Yuval N. *Homo Deus: a Brief History of Tomorrow*. New York, NY: Harper Perennial, 2018.

ENDNOTES

www.ingramcontent.com/pod-product-compliance
Lightning Source LLC
Chambersburg PA
CBHW072147100526
44589CB00015B/2129